Foundations for Algebra
Parent Guide for Years 1 and 2

Managing Editors:

Elizabeth Coyner
    Bear River School
Beverly Brockhoff
    Glen Edwards Middle School

Illustrator:

Jonathan Weast
    Weast & Weast Illustration Studio
    Sacramento, California

Technical Assistance:

Bethany Sorbello
    CPM Educational Program
Thu Pham
    The CRESS Center
    University of California, Davis

Developed by CPM Educational Program

Program Directors:

Judith Kysh
    Departments of Mathematics and Education
    San Francisco State University
Tom Sallee
    Department of Mathematics
    University of California, Davis
Brian Hoey
    CPM Educational Program

v 3.1

Credits for the First Edition

Heidi Ackley
Steve Ackley
Elizabeth Baker
Bev Brockhoff
Ellen Cafferata
Elizabeth Coyner
Scott Coyner
Sara Effenbeck
William Funkhouser

Brian Hoey
Judy Kysh
Kris Petersen
Robert Petersen
Edwin Reed
Stacy Rocklein
Kristie Sallee
Tom Sallee
Howard Webb

Technical Assistance

Jennifer Buddenhagen
Grace Chen
Janelle Petersen

David Trombly
Erika Wallender
Emily Wheelis

Introduction to the Parent Guide

Welcome to the *Foundations for Algebra Parent Guide.* The purpose of this guide is to assist you should your child need help with homework or the ideas in the course. We believe all students can be successful in mathematics as long as they are willing to work and ask for help when they need it. We encourage you to contact your child's teacher if your student has additional questions that this guide does not answer. **Assistance with most homework problems is available at www.hotmath.com.**

The guide is written for use with either Year 1 or Year 2 of the series. It is arranged by the topical bar graph displayed at the beginning of each chapter. These topics are: number sense, algebra and functions, measurement and geometry, statistics, data, and probability, and mathematical reasoning. Each topic is referenced to the chapter in which the major development of the concept occurs. Detailed examples follow a summary of the concept or skill and include complete solutions. The examples are similar to the work your child has done in class. Additional problems, with answers, are provided for your child to practice.

There will be some topics that your child understands quickly and some concepts that may take longer to master. The big ideas of the course take time to learn. This means that students are not necessarily expected to master a concept when it is first introduced. When a topic is first introduced in the textbook, there will be several problems to do for practice. Succeeding lessons and homework assignments will continue to practice the concept or skill over weeks and months so that mastery will develop over time.

Practice and discussion are required to understand mathematics. When your child comes to you with a question about a homework problem, often you may simply need to ask your child to read the problem and then ask her/him what the problem is asking. Reading the problem aloud is often more effective than reading it silently. When you are working problems together, have your child talk about the problems. Then have your child practice on his/her own.

Below is a list of additional questions to use when working with your child. These questions do not refer to any particular concept or topic. Some questions may or may not be appropriate for some problems.

- What have you tried? What steps did you take?

- What didn't work? Why didn't it work?

- What have you been doing in class or during this chapter that might be related to this problem?

- What does this word/phrase tell you?

- What do you know about this part of the problem?

- Explain what you know right now.

- What do you need to know to solve the problem?

- How did the members of your study team explain this problem in class?

- What important examples or ideas were highlighted by your teacher?

- Can you draw a diagram or sketch to help you?

- Which words are most important? Why?

- What is your guess/estimate/prediction?

- Is there a simpler, similar problem we can do first?

- How did you organize your information? Do you have a record of your work?

- Have you tried Guess and Check, making a list, looking for a pattern, etc.?

If your student has made a start at the problem, try these questions.

- What do you think comes next? Why?
- What is still left to be done?
- Is that the only possible answer?
- Is that answer reasonable?
- How could you check your work and your answer?
- How could your method work for other problems?

If you do not seem to be making any progress, you might try these questions.

- Let's look at your notebook, class notes, and Tool Kit. Do you have them?
- Were you listening to your team members and teacher in class?
  What did they say?
- Did you use the class time working on the assignment?
  Show me what you did.
- Were the other members of your team having difficulty with this as well?
  Can you call your study partner or someone from your study team?

This is certainly not a complete list; you will probably come up with some of your own questions as you work through the problems with your child. Ask any question at all, even if it seems too simple to you.

To be successful in mathematics, students need to develop the ability to reason mathematically. To do so, students need to think about what they already know and then connect this knowledge to the new ideas they are learning. Many students are not used to the idea that what they learned yesterday or last week will be connected to today's lesson. Too often students do not have to do much thinking in school because they are usually just told what to do. When students understand that connecting prior learning to new ideas is a normal part of their education, they will be more successful in this mathematics course (and any other course, for that matter). The student's responsibilities for learning mathematics include the following:

- Actively contributing in whole class and study team work and discussion.
- Completing (or at least attempting) all assigned problems and turning in assignments in a timely manner.
- Checking and correcting problems on assignments (usually with their study partner or study team) based on answers and solutions provided in class.
- Studying the tutorial solutions to homework problems available at www.hotmath.com.
- Asking for help when needed from his or her study partner, study team, and/or teacher.
- Attempting to provide help when asked by other students.
- Taking notes and using his/her Tool Kit when recommended by the teacher or the text.
- Keeping a well-organized notebook.
- Not distracting other students from the opportunity to learn.

Assisting your child to understand and accept these responsibilities will help him or her to be successful in this course, develop mathematical reasoning, and form habits that will help her/him become a life-long learner.

Table of Contents by Strand and Topic

Table of Contents by Course: Year 1

Table of Contents by Course: Year 2

Number Sense

Integers

Absolute Value

Ratio

Proportions

Percentage of Increase and Decrease

Simple and Compound Interest

Equivalent Fractions

Addition and Subtraction of Fractions

Multiplying and Dividing Fractions

Fraction, Decimal, and Percent Equivalents

Laws of Exponents

Scientific Notation

Diamond Problems

MODELING INTEGERS

Integers are positive and negative whole numbers and zero. A negative integer is written with a negative sign in front of the numeral, such as -6 and -11, and is to the left of zero on the number line. A positive integer may be written with or without a positive sign in front of the number. For example, 5 and +5 are both ways of indicating 5 units (or "steps") to the right of zero on the number line.

We use tiles to model integers. Positive integer tiles are represented with a positive sign (+). Negative integer tiles are represented with a negative sign (–). Zero is represented by an equal number of positive signs and negative signs and is referred to as a neutral field composed of zero pairs. Zero pairs are circled.

4 is modeled as   ++++   and recorded as + 4.

-3 is modeled as   – – –   and recorded as -3.

0 is modeled as   (+/–)   and recorded as 0. (A negative neutralizes a positive or makes it zero.)

(+/–)(+/–)(+/–) +   This model represents +1 because any number of zeros does not change the value of the number being modeled.

For additional information, see Year 1, Chapter 2, problems GS-5 and 6 on pages 35-36 or Year 2, Chapter 2, problem FT-29 on page 49.

Examples

a)   zero pairs in a neutral field   b)   zero pairs in the number 4   c)   -3 with more than 6 tiles

(+/–)(+/–)(+/–)(+/–)   (+/–)(+/–)(+/–) + + + +   (+/–)(+/–)(+/–) – – –

Problems

Represent the following integers with at least 9 tiles.

1.   -4   2.   5   3.   -6   4.   0

Answers

Note: In this section of the guide, answers are provided as examples of correct answers. There may be several acceptable models for some problems.

1. (+/–)(+/–)(+/–) – – – –   2. (+/–)(+/–) + + + + +   3. (+/–)(+/–) – – – – – –   4. (+/–)(+/–)(+/–)(+/–)(+/–)

ADDITION OF INTEGERS WITH TILES

Tiles can be used to model addition of integers.

4 + (-3)

Draw a box.

Build the first number.
Put it in the box.

Build the second number.
Put it in the box.

Circle all the zero pairs.

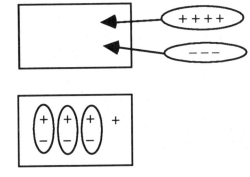

Count the uncircled tiles and record the equation showing the sum: 4 + (-3) = 1

For more information on modeling addition of integers, see Year 1, Chapter 2, problems GS-5 and 14 or Year 2, Chapter 2, problems FT-29 and 30.

Example 1

Draw tiles to compute -3 + (-2).

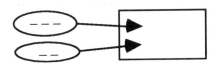    -3 + (-2) = -5

There are no zero pairs here to circle, so the answer is -5.

Example 2

Draw tiles to add 5 + (-6). Remember to circle the zero pairs.

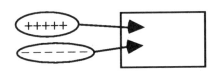 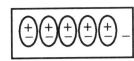   5 + (-6) = -1

Example 3

Draw tiles to compute -3 + 3.

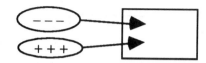   -3 + 3 = 0

Problems

Draw tiles to compute the following problems.

1. -2 + (-3)   2. -2 + (-5)   3. 5 + 2   4. 4 + (-4)

5. 5 + (-3)   6. -5 + 3   7. -3 + 7   8. -5 + 6

9. -2 + 2   10. 1 + (-4) + (-1)

Answers

1. -2 + (-3) = -5   2. -2 + (-5) = -7   3. 5 + 2 = 7
    − − − − −   − − − − − −   + + + + + + +

4. 4 + (-4) = 0   5. 5 + (-3) = 2   6. -5 + 3 = -2
    (+)(+)(+)(+)   (+)(+)(+) + +   (+)(+)(+) − −
    ⊖ ⊖ ⊖ ⊖   ⊖ ⊖ ⊖   ⊖ ⊖ ⊖

7. -3 + 7 = 4   8. -5 + 6 = 1   9. -2 + 2 = 0
    (+)(+)(+) + + + +   (+)(+)(+)(+)(+) +   (+)(+)
    ⊖ ⊖ ⊖   ⊖ ⊖ ⊖ ⊖ ⊖   ⊖ ⊖

10. 1 + (-4) + (-1)= -4
     (+) − − − −
     ⊖

ADDITION OF INTEGERS IN GENERAL

When integers are added using the tile model, zero pairs are only formed if the two numbers have different signs. After circling the zero pairs, count the uncircled tiles to find the sum. If the signs are the same, no zero pairs are formed. Count the tiles and the result is the sum of the tiles. Integers can be added without building models using the rules below.

- If the signs are the same, add the numbers and keep the same sign.

- If the signs are different, ignore the signs (that is, use the absolute value of each number.) Subtract the number closest to zero from the number farthest from zero. The sign of the answer is the same as the number that is farthest from zero, that is, the number with the greater absolute value.

For more information on the rules for addition of integers, see Year 2, Chapter 2, problem FT-50 on page 55.

Example

-4 + 2   (-4) is farther from zero on the number line than 2, so $|-4| - |2| = 4 - 2 = 2$. The answer is -2, since the "4" is negative.

Problems

Use either the Tile (Neutral Field) Model or the rules above to find these sums and/or differences.

1. 4 + (-2)
2. 6 + (-1)
3. 7 + (-7)
4. -10 + 6

5. -8 + 2
6. -12 + 7
7. -5 + (-8)
8. -10 + (-2)

9. -11 + (-16)
10. -8 + 10
11. -7 + 15
12. -26 + 12

13. -3 + 4 + 6
14. 56 + 17
15. 7 + (-10) + (-3)
16. -95 + 26

17. 35 + (-6) + 8
18. -113 + 274
19. 105 + (-65) + 20
20. -6 + 2 + (-4) + 3 + 5

21. 5 + (-3) + (-2) + (-8)
22. -6 + (-3) + (-2) + 9
23. -6 + (-3) + 9

24. 20 + (-70)
25. 12 + (-7) + (-8) + 4 + (-3)
26. -26 + (-13)

27. -16 + (-8) + 9
28. 12 + (-13) + 18 + (-16)

29. 50 + (-70) + 30
30. 19 + (-14) + (-5) + 20

Answers

1. 2
2. 5
3. 0
4. -4
5. -6
6. -5

7. -13
8. -12
9. -27
10. 2
11. 8
12. -14

13. 7
14. 73
15. -6
16. -69
17. 37
18. 161

19. 60
20. 0
21. -8
22. -2
23. 0
24. -50

25. -2
26. -39
27. -15
28. 1
29. 10
30. 20

SUBTRACTION OF INTEGERS

Tiles are also useful to subtract integers. In order to do so, first build a neutral field using zero pairs. Then remove tiles--positive or negative as indicated by the sign of the number that is subtracted--from the neutral field. For example: 0 - 4

Start with a neutral field of zero pairs.

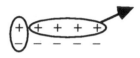

Remove 4 positive tiles. What is left?

The zero pairs left are circled and the remaining tiles are counted. The answer is -4. Record the difference in an equation.

$$0 - 4 = -4$$

In $-2 - (-3)$, build the first integer with a neutral field.

Next, remove the second integer.

Be careful to circle all remaining zero pairs.

Record the difference in an equation.

$$-2 - (-3) = 1$$

Sometimes a neutral field is not needed. For example, $-6 - (-3)$ does not need a neutral field.

Build the first integer.

Ask, "Are there enough negative tiles so that I can remove 3 negatives?" If the answer is "yes," a neutral field is unnecessary. Remove the three negative tiles and record the answer in an equation.

$$-6 - (-3) = -3$$

Be careful with problems like $-6 - 3$. This problem means -6 minus a positive 3. Build a neutral field because otherwise $-\ -\ -\ -\ -\ -\ (-6)$ has no positive tiles to remove.

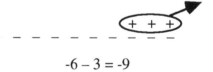

$$-6 - 3 = -9$$

For more information on modeling subtraction of integers, see Year 1, Chapter 3, problems PR-15 through 16 on pages 72-73 and PR-41 through 43 and PR-50 on pages 79-80 or Year 2, Chapter 2, problems FT-55 through 58 on pages 57-58.

Problems

Find each difference. Use tiles for at least the first five problems.

1. -6 – (-2)   2. 5 – (-3)   3. 6 – (-3)

4. -7 – 3   5. 7 – (-3)   6. 7 – 3

7. 5 – (3)   8. -12 – (-10)   9. -12 – 10

10. 12 – (-10)   11. -6 – (-3) – 5   12. 6 – (-3) – 5

13. 8 – (-8)   14. -9 – 9   15. -9 – 9 – (-9)

Answers

1. -4

2. 8

3. 9

4. -10

5. 10

6. 4   7. 2

8. -2   9. -22   10. 22   11. -8

12. 4   13. 16   14. -18   15. -9

In the next six examples, compare (a) to (b), (c) to (d), and (e) to (f). Notice that examples (a), (c), and (e) are subtraction problems and examples (b), (d), and (f) are addition problems. The answers to each pair of examples are the same. Also notice that the second integers in the pairs are opposites (that is, they are the same distance from zero on opposite sides of the number line) while the first integers in each pair are the same.

a)   2 – (-6)   2 – (-6) = 8

b)   2 + 6   2 + 6 = 8

c)   -3 – (-4)   -3 – (-4) = 1

d)   -3 + 4   -3 + 4 = 1

e)   -4 – (-3)   -4 – (-3) = -1

f)   -4 + 3   -4 + 3 = -1

Subtracting an integer is the same as adding its opposite. This is summarized below.

Subtraction of Integers in General: To find the difference of two integers, change the subtraction sign to an addition sign. Next change the sign of the integer you are subtracting, then apply the rules for addition of integers.

For more information on the rules for subtracting integers, see Year 1, Chapter 3, problem PR-50 on page 82 or Year 2, Chapter 2, problems FT-86 through 88 on pages 66-67.

Examples

Use the rule for subtracting integers stated above to compute.

a)   9 – (-12) becomes 9 + (+12) = 21   b)   -9 – (-12) becomes -9 + (+12) = 3

c)   -9 – 12 becomes -9 + (-12) = -21   d)   9 – 12 becomes 9 + (-12) = -3

Problems

Use the rule stated above to find each difference.

1. 9 – (-3)
2. 9 – 3
3. -9 – 3
4. -9 – (-3)
5. -14 – 15
6. -16 – (-15)
7. -40 – 62
8. -40 – (-62)
9. 40 – 62
10. 40 – (-62)
11. -5 – (-3) – 5 – 6
12. -5 – 3 – (-5) – (-6)
13. 5 – 3 – (-5) – 6
14. 5 – (-4) – 6 – (-7)
15. -125 – (-125) – 125
16. 5 – (-6)
17. 12 – 14
18. 20 – 25
19. -3 – 2
20. -7 – 3
21. -10 – 5
22. -30 – 7
23. -3 – (-3)
24. -3 – (-4)
25. 10 – (-3)
26. 5 – (-9)
27. 27 – (-7)
28. 15 – 32
29. -58 – 37
30. -79 – (-32)
31. -62 – 81
32. -106 – 242
33. 47 – (-55)
34. 257 – 349
35. -1010 – (-1010)

Answers

1. 12
2. 6
3. -12
4. -6
5. -29
6. -1
7. -102
8. 22
9. -22
10. 102
11. -13
12. 3
13. 1
14. 10
15. -125
16. 11
17. -2
18. -5
19. -5
20. -10
21. -15
22. -37
23. 0
24. 1
25. 13
26. 14
27. 34
28. -17
29. -95
30. -47
31. -143
32. -348
33. 102
34. -92
35. 0

MULTIPLICATION OF INTEGERS WITH TILES

Multiplication is repeated addition. $4 \cdot 3$ is the same as $3 + 3 + 3 + 3$. Integer tiles can be used to model multiplication of integers.

Modeling Multiplication With Tiles When the First Factor is Positive: An example is $2(-4)$. The first factor (2) is a positive number and means that two groups will go into a box. The second factor (-4) tells how many are in each group. In this example, four negatives are in each group.

Step 1:   Start with an empty box. Build two sets of four negative tiles.

Step 2:   Physically push the tiles into the box, one group at a time, to show the repeated addition.

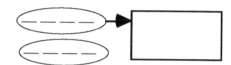

    a) "One group of -4 ..."

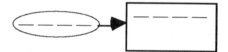

    b) "Two groups of -4."

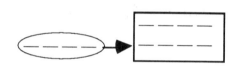

Step 3:   Count the tiles in the box to find the product.

Step 4:   Write the multiplication equation.

$$2(-4) = -8$$

For more information on modeling multiplication of integers, see Year 1, Chapter 2, problem GS-48 on page 49 and Chapter 3, problem PR-54 on page 83 and PR-71 on page 88 or Year 2, Chapter 2, problems FT-102 through 107 on pages 70-72.

Problems

Use the tile drawings to multiply in the first five problems below. Remember to write an equation to show the answer. Find the product for each of the remaining problems.

1. 3(-4)   2. 2(-6)   3. 5(-1)   4. 6(-2)   5. 7(-3)

6. (8)(-7)   7. 7(-12)   8. 13(-5)   9. (9)(-8)   10. 7(-5)

11. 8(-6)   12. 9(-7)   13. 10(8)   14. 22(-10)   15. 4(-30)

Answers

1. $3(-4) = -12$

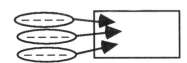

2. $2(-6) = -12$

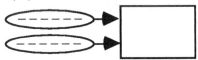

3. $5(-1) = -5$

4. $6(-2) = -12$

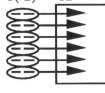

5. $7(-3) = -21$

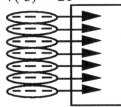

6. -56   7. -84   8. -65   9. -72   10. -35

11. -48   12. -63   13. 80   14. -220   15. -120

Modeling Multiplication With Tiles When the First Factor is Negative: The previous example modeled multiplying integers like 3(-2), which means adding in 3 groups of -2 to get a total of -6. When the first number is a negative, such as -2(3), the multiplication problem becomes repeated subtraction as shown in Example 1 below.

Example 1

Use tile drawings to multiply (-2)(3).

Step 1:   Build a large neutral field.

+ + + + + + +
− − − − − − −

Step 2:   Circle and remove one group of +3. Do it
again. You removed two groups of +3.

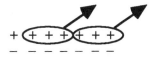

Step 3:   Circle any zero pairs. Count the remaining tiles.

-6

Step 4:   Record the equation, showing the product.

$(-2)(3) = -6$

11

Example 2

Use tile drawings to multiply (-3)(-3).

Step 1:　　+ + + + + + + + +

　　　　　　－ － － － － － － － －

Step 2:　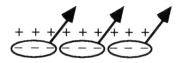

Step 3:　　+ + + + + + + + +　　　9

Step 4:　　(-3)(-3) = 9

MULTIPLICATION OF INTEGERS IN GENERAL

To multiply without tiles, use the absolute value of the product and use the rules below to determine the sign of the answer.

- When pairs of integers with the same sign are multiplied, the product is positive.

- When pairs of integers with different signs are multiplied, the product is negative.

For more information about the rules for multiplication of integers, see Year 1, Chapter 3, problems PR-66 on page 86 and PR-71 on page 88 or Year 2, Chapter 2, problems FT-105 through 108 on pages 71-72.

Problems

Use either the Tile (Neutral Field) Model or the rules above to find each product.

1. (-3)(2)　　　　2. (-5)(4)　　　　3. (-11)(7)　　　　4. (-12)(5)

5. (-23)(7)　　　6. (4)(-6)　　　　7. (7)(-9)　　　　8. (15)(-8)

9. (35)(-3)　　　10. (115)(-7)　　11. (-5)(-6)　　　12. (-7)(-8)

13. (-24)(-8)　　14. (-137)(-4)　　15. (-231)(-17)　　16. (-3)(5)(-6)

17. (-7)(-4)(3)　　18. (-5)(-5)(-5)　　19. (-3)(-2)(6)(4)　　20. (-5)(-4)(-2)(-3)

21. -2(4)　　　　22. -5(1)　　　　23. 2(-3)　　　　24. -15(-10)

25. -15(10)　　　26. 12(-12)　　　27. -11(-10)　　　28. (-2)(-3)(-5)

29. (-2)(-3)(4)(6)　　30. (-2)(-3)(-4)(-6)　　31. (-2)(-3)(4)(-6)　　32. 2(-3)(4)(-6)

33. -2(3)(-5)

Answers

1. -6	2. -20	3. -77	4. -60	5. -161
6. -24	7. -63	8. -120	9. -105	10. -805
11. 30	12. 56	13. 192	14. 548	15. 3927
16. 90	17. 84	18. -125	19. 144	20. 120
21. -8	22. -5	23. -6	24. 150	25. -150
26. -144	27. 110	28. -30	29. 144	30. 144
31. -144	32. 144	33. 30		

DIVISION OF INTEGERS

When dividing integers, the rules for the sign of the product are the same as those for multiplication. Use the absolute values to divide and then determine the sign. When dividing two integers with the same sign, the result is positive. When dividing two integers with different signs, the result is negative.

Examples $14 \div (-7) = -2$ $-14 \div (-7) = 2$

For more information on division of integers, see Year 1, Chapter 6, problems MB-109 and 110 on pages 210-11 or Year 2, Chapter 2, problems FT-118 through 121 on pages 74-75.

Problems

Divide.

1. $10 \div (-2)$ 2. $15 \div (-3)$ 3. $108 \div (-3)$

4. $258 \div (-6)$ 5. $-14 \div 7$ 6. $-56 \div 4$

7. $-110 \div 11$ 8. $-95 \div 95$ 9. $-68 \div (-4)$

10. $-125 \div (-25)$ 11. $-96 \div (-12)$ 12. $-115 \div 23$

13. $342 \div (-6)$ 14. $-217 \div (-217)$ 15. $-2088 \div (-24)$

Answers

1.   -5   2.   -5   3.   -36   4.   -43   5.   -2

6.   -14   7.   -10   8.   -1   9.   17   10. 5

11. 8   12. -5   13. -57   14. 1   15. 87

ABSOLUTE VALUE

Absolute value is the distance a point is from zero on the number line. It is always a positive number since it measures the physical distance from zero.

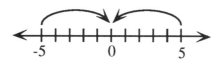

The symbol for absolute value is | |. On the number line above, both 5 and -5 are 5 units from zero. This distance is displayed as |-5| = 5 and is read, "The absolute value of negative five equals five." Similarly, |5| = 5 means, "The absolute value of five is five."

|x| = 5 means that x could be <u>either</u> 5 or -5 because both of those points are five units from zero.

The problem |x| = -5 has no solution because the absolute value of a number has to be positive. The only exception to this is when a negative sign appears outside the absolute value symbol.

For additional information, see Year 1, Chapter 2, problem GS-76 on page 56 or Year 2, Chapter 2, problem FT-44 on page 53.

Examples

a) |-6| = 6

b) |7| = 7

c) - |-5| = -5

d) |x| = -3

e) |x| = 9 $\Rightarrow$ x = -9 or 9

f) - |x| = -3 $\Rightarrow$ x = -3 or 3

g) |3 - 8| = |-5| = 5

Part (d) has no solution, since any absolute value is positive.

Notice the negative sign outside the absolute value symbol in examples (c) and (f). This sign means "the opposite of the absolute value."

Problems

Determine the absolute value or the values of x.

1. |-11|

2. |12|

3. |x| = 4

4. |x| = 16

5. |x| = 24

6. |x| = 13

7. |-9|

8. |x| = -13

9. - |x| = -13

10. - |7|

11. |x| = 7

12. |-7|

13. |5 – 8|

14. |-6 – 3|

15. |-6 + 3|

Answers

1. 11

2. 12

3. 4, -4

4. 16, -16

5. 24, -24

6. 13, -13

7. 9

8. no solution

9. 13, -13

10. -7

11. 7, -7

12. 7

13. 3

14. 9

15. 3

RATIO

A **ratio** is a comparison of two quantities by division. It can be written in several ways:

$$\frac{65 \text{ miles}}{1 \text{ hour}}, \; 65 \text{ miles}: 1 \text{ hour}, \text{ or } 65 \text{ miles to 1 hour.}$$

Both quantities of a ratio can be multiplied by the same number. Use a **ratio table** to organize the multiples. Each ratio in the table will be equivalent to the others. Patterns in the ratio table can be used in problem solving.

For additional information, see Year 1, Chapter 5, problem GO-1 on page 145 or Year 2, Chapter 6, problem RS-3 on page 208.

Example 1

120 cups of coffee can be made from one pound of coffee beans. Doubling the amount of coffee beans will double the number of cups of coffee that can be made. Use the doubling pattern to complete the ratio table for different weights of coffee beans.

Pounds of coffee beans	1	2	4	8
Cups of coffee	120	240		

Doubling two pounds of beans doubles the number of cups of coffee made, so for 4 pounds of beans, $2 \cdot 240 = 480$ cups of coffee are made. Since 4 pounds of beans make 480 cups of coffee, 8 pounds of beans make $2 \cdot 480 = 960$ cups of coffee.

Example 2

You can use the ratio table from Example 1 to determine how many cups of coffee you could make from 6 pounds of beans. Add another column to your ratio table.

Pounds of coffee beans	1	2	4	8	6
Cups of coffee	120	240	480	960	

You know that the value of 6 is halfway between the values of 4 and 8. The value halfway between 480 and 960 is 720, so 6 pounds of beans should make 720 cups of coffee.

Example 3

Jane's cookie recipe uses $2\frac{1}{2}$ cups of flour and 2 eggs. She needs to know how much flour she will need if she uses 9 eggs. Jane started a ratio table but discovered that she could not simply keep doubling because 9 is not a multiple of 2. She used other patterns to find her answer. Study the top row of Jane's table to find the patterns she used, then complete the table.

Number of eggs	2	4	12	36	9
Cups of flour	$2\frac{1}{2}$				

Jane doubled to get 4, tripled 4 to get 12, and tripled again to get 36. Then she divided 36 by four to get to 9. She followed the same steps to complete the "cups of flour" row. Jane will need $11\frac{1}{4}$ cups of flour.

Number of eggs	2	4	12	36	9
Cups of flour	$2\frac{1}{2}$	5	15	45	$11\frac{1}{4}$

> In the examples above, the ratio tables have the ratios listed in the order in which they were calculated. When the values in a row of a ratio table are provided, it is acceptable to complete the ratio table in the order that fits the patterns in the table. In most cases it is easier to see patterns by listing the values in the first (or top) row in order.

Example 4

Peter and Mandy each completed this ratio table correctly, but they did it in a different order using different patterns.

2	4	6	8
17			

Peter tripled $\frac{2}{17}$ to get $\frac{6}{51}$. Then he doubled $\frac{2}{17}$ to get $\frac{4}{34}$ and doubled that to get $\frac{8}{68}$. Mandy, however, doubled $\frac{2}{17}$ to get $\frac{4}{34}$ and doubled that to get $\frac{8}{68}$. Then she used the halfway values to get $\frac{6}{51}$.

Problems

Complete each ratio table.

1.

2	4	6	8	10	12	16	20
5	10						

2.

3	6	9	12	15	18	21	30
7	14						

3.

5			20	30	45	60	100
2	4	6					

4.

7	14	35			49		700	
4			40		100			1000

5.

6	12	24	18			48		120
11				66			99	

6.

6	12	18	30			360	
5.7				57	114		684

7.

9	36	45			180	900	300
48			120	480			

8.

10	15	20	25		35		4000
8.5				34		340	

Answers

1.

2	4	6	8	10	12	16	20
5	10	15	20	25	30	40	50

2.

3	6	9	12	15	18	21	30
7	14	21	28	35	42	49	70

3.

5	10	15	20	30	45	60	100
2	4	6	8	12	18	24	40

4.

7	14	35	70	49	175	700	1750
4	8	20	40	28	100	400	1000

5.

6	12	24	18	36	48	54	120
11	22	44	33	66	88	99	220

6.

6	12	18	30	60	120	360	720
5.7	11.4	17.1	28.5	57	114	342	684

7.

9	36	45	22.5	90	180	900	300
48	192	240	120	480	960	4800	1600

8.

10	15	20	25	40	35	400	4000
8.5	12.75	17	21.25	34	29.75	340	3400

PROPORTIONS

SOLVING PROPORTIONS

A **proportion** is an equation that states that two ratios are equal.

For example: $\frac{3}{8} = \frac{6}{16}$, $\frac{4}{12} = \frac{6}{18}$, and $\frac{2}{5} = \frac{x}{40}$.

Some proportions have an unknown value such as the x in the last example. Because the unknown value is part of two equivalent fractions, there are several methods to find the value of x in a proportion like $\frac{2}{5} = \frac{x}{40}$.

One method of solving a proportion is to use a ratio table like the one at right. The ratio table shows that $x = 16$.

2	4	8	16
5	10	20	40

Another method is to use the Giant **1**, as shown at right.

$$\frac{2}{5} = \frac{x}{40} \Rightarrow \frac{2}{5} \cdot \frac{8}{8} = \frac{16}{40} \Rightarrow x = 16$$

For additional information, see Year 1, Chapter 5, problem GO-80 on page 166 and Chapter 6, problem MB-8 on page 183 or Year 2, Chapter 6, problems RS-17 on page 211 and RS-23 on page 213.

Problems

Use a ratio table to solve each proportion, then use the Giant **1** to solve them again.

1. $\frac{3}{7} = \frac{x}{56}$

2. $\frac{4}{9} = \frac{x}{72}$

3. $\frac{2}{5} = \frac{x}{35}$

4. $\frac{2}{11} = \frac{x}{88}$

5. $\frac{14}{17} = \frac{x}{85}$

6. $\frac{12}{19} = \frac{108}{x}$

Answers

1. $x = 24$ 2. $x = 32$ 3. $x = 14$ 4. $x = 16$ 5. $x = 70$ 6. $x = 171$

APPLICATIONS USING PROPORTIONS

Proportions can be used to solve various types of problems.

To solve a problem using a proportion, write the known information in one ratio and the unknown value as part of another ratio with a variable to represent what is unknown. Keep the units in each ratio in the same order, such as $\frac{65 \text{ miles}}{1 \text{ hour}} = \frac{x \text{ miles}}{4 \text{ hours}}$. Solve the proportion using method.

Example 1

Balvina knows that 6 cups of rice will make enough Spanish rice to feed 15 people. She needs to know how many cups of rice are needed to feed 135 people. Write a proportion to solve the problem.

What we know is $\frac{6 \text{ cups}}{15 \text{ people}}$, and what we want to know is $\frac{x \text{ cups}}{135 \text{ people}}$, so $\frac{6}{15} = \frac{x}{135}$.

We can solve the proportion using a ratio table.

Cups of Rice	6	12	18	24	48	54
Servings	15	30	45	60	120	135

Balvina needs 54 cups of rice to feed 135 people.

We can also solve the proportion using the Giant **1**.

$\frac{6}{15} \cdot \boxed{1} = \frac{x}{135}$ Since $135 \div 15 = 9$, use $\frac{9}{9}$ in the Giant **1**.

$\frac{6}{15} \cdot \boxed{\frac{9}{9}} = \frac{54}{135}$ Again, Balvina needs 54 cups of rice.

Example 2

Ivanna needs to buy 360 cherries for a large salad. She can buy 9 cherries for $0.57. How much will 360 cherries cost Ivanna? Write a proportion to solve the problem.

What we know is $\frac{9 \text{ cherries}}{\$0.57}$, and what we want to know is $\frac{360 \text{ cherries}}{\$x}$, so $\frac{9}{0.57} = \frac{360}{x}$.

We can solve the proportion using a ratio table. To get from 9 cherries to 360 cherries, we multiply the known ratio by 10, then by 4.

Number of Cherries	9	90	360
Cost	$0.57	$5.70	$22.80

The 360 cherries will cost Ivanna $22.80.

We can also solve the proportion using the Giant **1**.

$$\frac{9}{0.57} \cdot \boxed{} = \frac{360}{x}$$ Since $360 \div 9 = 40$, use $\frac{40}{40}$ in the Giant **1**.

$$\frac{9}{0.57} \cdot \boxed{\frac{40}{40}} = \frac{360}{22.80}$$ Again, the 360 cherries will cost Ivanna $22.80.

The problems below involve rate. For additional information, see Year 2, Chapter 9, problems CB-13 through 17 on pages 335-37.

Example 3

Elaine can plant 6 flowers in 15 minutes. How long will it take her to plant 30 flowers at the same rate? Write a proportion to solve the problem.

What we know is $\frac{6 \text{ flowers}}{15 \text{ minutes}}$, and what we want to know is $\frac{30 \text{ flowers}}{x \text{ minutes}}$, so $\frac{6}{15} = \frac{30}{x}$.

We can solve the proportion using a ratio table.

Number of Flowers	6	12	18	24	30
Number of Minutes	15	30	45	60	75

It will take Elaine 75 minutes (or 1 hour and 15 minutes) to plant 30 flowers.

We can also solve the proportion using the Giant **1**.

$$\frac{6}{15} \cdot \boxed{} = \frac{30}{x}$$ Since $30 \div 6 = 5$, use $\frac{5}{5}$ in the Giant **1**.

$$\frac{6}{15} \cdot \boxed{\frac{5}{5}} = \frac{30}{75}$$ Again, it will take Elaine 75 minutes to plant 30 flowers.

Problems

Solve the following proportions involving rate.

1.   A plane travels 3400 miles in 8 hours. How far would it travel in 6 hours at this rate?

2.   Shane rode his bike for 2 hours and traveled 12 miles. At this rate, how long would it take him to travel 22 miles?

3.   Selina's car used 15.6 gallons of gas to go 234 miles. At this rate, how many gallons would it take her to go 480 miles?

Answers

1. 2550 miles

2. $3\frac{2}{3}$ hours

3. 32 gallons

Proportions can also be used to solve percent problems by writing $\frac{\%}{100} = \frac{part}{whole}$. For more information, refer to Year 1, Chapter 6, problem MB-64 on page 199 or Year 2, Chapter 6, problem RS-37 on page 217.

Example 4

32 is what percent of 40?

$\frac{32}{40} = \frac{x}{100}$

x = 80, so 32 is 80% of 40.

Example 5

15% of 80 is what number?

$\frac{15}{100} = \frac{x}{80}$

x = 12

Problems

Solve the following percent problems using proportions.

1. What is 8% tax on an $80 purchase?

2. Bev scored 33 points on a 40-point quiz. What was her percent correct?

3. Elizabeth answered 114 questions correctly on her science test. Her score was 95%. How many questions were on the test?

Answers

1. $6.40

2. 82.5%

3. 120 questions

The ratio of pairs of corresponding sides of similar figures are equal. When given a pair of similar figures, you can use a proportion to find the length of an unknown side. For more information, refer to Year 1, Chapter 6, problems MB-87 through 90 on pages 205-06 or Year 2, Chapter 6, problems RS-75 through 77 on pages 229-30.

Example 6

Two similar triangles.

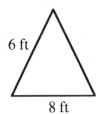 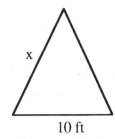

$$\frac{x \text{ ft}}{6 \text{ ft}} = \frac{10 \text{ ft}}{8 \text{ ft}} \qquad x = 7.5 \text{ ft}$$

Example 7

Two similar pentagons.

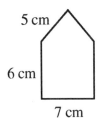 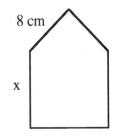

$$\frac{5 \text{ cm}}{8 \text{ cm}} = \frac{6 \text{ cm}}{x \text{ cm}} \qquad x = 9.6 \text{ cm}$$

Problems

Solve for the unknown side in these similar figures. Round answers to the nearest tenth.

1.

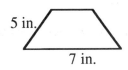

2.

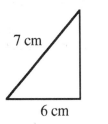 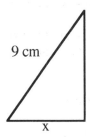

3.

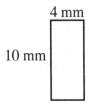

4.

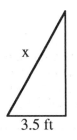 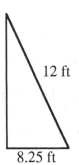

Answers

1. 11.4 in.   2. 7.7 cm   3. 10 mm   4. 5.1 ft.

PERCENTAGE OF INCREASE AND DECREASE

Another kind of problem that can be solved using proportions involves finding the percentage of increase or decrease. Use the following formula to find percent:

$$\frac{\%}{100} = \frac{\text{part}}{\text{whole}}$$

When change in percent is found, the same concept is used. The formula becomes:

$$\frac{\%}{100} = \frac{\text{change (increase or decrease)}}{\text{original amount}}$$

For additional information, see Year 2, Chapter 7, problem CT-105 on page 277.

Example 1

A town's population grew from 1,879 to 7,426 over five years. What was the percentage of increase?

- Subtract to find the change:

$$7426 - 1879 = 5547$$

- Put the known numbers in the proportion:

$$\frac{\%}{100} = \frac{5547}{1879} \begin{array}{l}\text{increase}\\\text{original}\end{array}$$

- The percentage becomes x, the unknown:

$$\frac{x}{100} = \frac{5547}{1879}$$

- Cross multiply (see "ratio equations," p. 81):

$$1879x = 554{,}700$$

- Divide each side by 1879:

$$x = 295.2\% \text{ increase}$$

The population increased by 295.2%

Example 2

A Sumo wrestler retired from Sumo wrestling and went on a diet. When he retired he weighed 385 pounds. After two years he weighed 238 pounds. What was the percentage of decrease in his weight?

- Subtract to find the change:

$$385 - 238 = 147$$

- Put the known numbers in the proportion:

$$\frac{\%}{100} = \frac{147}{385} \begin{array}{l}\text{decrease}\\\text{original}\end{array}$$

- The percentage becomes x, the unknown:

$$\frac{x}{100} = \frac{147}{385}$$

- Cross multiply (see "ratio equations," p. 81):

$$385x = 14{,}700$$

- Divide each side by 385:

$$x = 38.\overline{18}\%.$$

His weight decreased by $38.\overline{18}\%$.

Problems

Solve the following problems.

1. Thirty years ago gasoline cost $0.30 per gallon on average. Today gasoline averages about $1.50 per gallon. What is the percentage of increase in the cost of gasoline?

2. When Spencer was 5, he was 28 inches tall. Today he is 5 feet 3 inches tall. What is the percentage of increase in Spencer's height?

3. The cars of the early 1900s cost $500. Today a new car costs an average of $27,000. What is the percentage of increase of the cost of an automobile?

4. The population of the U.S. at the first census in 1790 was 3,929,000 people. By 2000 the population had increased to 284,000,000! What is the percentage of increase?

5. In 2000 the rate for a first class U.S. postage stamp increased to $0.34. This represents a $0.31 increase since 1917. What is the percentage of increase since 1917?

6. In 1906 Americans consumed an average of 26.85 gallons of whole milk per year. By 1998 the average consumption was 8.32 gallons. What is the percentage of decrease in consumption of whole milk?

7. In 1984 there were 125 students for each computer in U.S. public schools. By 1998 there were 6.1 students for each computer. What is the percentage of decrease in the ratio of students to computers?

8. Sara bought a dress on sale for $30. She saved 45%. What was the original cost?

9. Pat was shopping and found a jacket with the original price of $120 on sale for $9.99. What was the percentage of decrease?

10. The price of a pair of pants decreased from $49.99 to $19.95. What was the percentage of decrease?

Answers

1. $1.50 - 0.30 = 1.20$; $\frac{x}{100} = \frac{1.20}{0.30}$; $x = 400\%$

2. $63 - 28 = 35$; $\frac{x}{100} = \frac{35}{28}$; $x = 125\%$

3. $27{,}000 - 500 = 26{,}500$; $\frac{x}{100} = \frac{26\,500}{500}$; $x = 5300\%$

4. $284{,}000{,}000 - 3{,}929{,}000 = 280{,}071{,}000$; $\frac{x}{100} = \frac{280071000}{3929000}$; $x = 7{,}128.3\%$

5. $0.34 - 0.31 = 0.03$; $\frac{x}{100} = \frac{0.31}{0.03}$; $x = 1033.3\%$

6. $26.85 - 8.32 = 18.53$; $\frac{x}{100} = \frac{18.53}{26.85}$; $x = 69.01\%$

7. $125 - 6.1 = 118.9$; $\frac{x}{100} = \frac{118.9}{125}$; $x = 95.12\%$

8. $100\% - 45\% = 55\%$; $\frac{55}{100} = \frac{30}{x}$; $x = \$54.55$

9. $120 - 9.99 = 110.01$; $\frac{x}{100} = \frac{110.01}{120}$; $x = 91.7\%$

10. $49.99 - 19.95 = 30.04$; $\frac{x}{100} = \frac{30.04}{49.99}$; $x = 60.1\%$

The idea of earning interest is introduced in Chapter 7 of Year 2. To find simple interest, use the formula $I = Prt$, where I = interest earned, P = principal, r = rate of interest, and t = time. For additional information, see Year 2, Chapter 7, problem CT-57 on page 265.

Example 1

Wayne earns 5.3% simple interest for 5 years on $3000. How much interest does he earn?

Put the numbers in the formula $I = Prt$.	$I = 3000(5.3\%)5$
Change the percent to a decimal.	$= 3000(0.053)5$
Multiply.	$= 795$ Wayne would earn $795 interest.

To find compound interest, use the formula $A = P(1 + r)^t$, where A = amount of money at the end of t years, P = principal, r = rate, and t = time. For additional information, see Year 2, Chapter 7, problem CT-58 on page 265.

Example 2

Use the numbers in Example 1 to find how much money Wayne would have if he earned 5.3% interest compounded annually.

Put the numbers in the formula $A = P(1 + r)^t$.	$A = 3000(1 + 5.3\%)^5$
Change the percent to a decimal.	$= 3000(1 + 0.053)^5$
Multiply.	$= 3883.86$

Wayne would have $3883.86.

To compare the difference in earnings when an amount is earning simple or compound interest, subtract the simple interest from the compound interest earned on the same amount of principal In these examples, Wayne would have $88.86 more with compound interest than he would have with simple interest: $3883.86 – ($3,000 + $795) = $88.86. Note that if the difference is negative, then simple interest earns more.

Problems

Solve the following problems.

1. Tong loaned Jody $50 for a month. He charged 5% simple interest for the month. How much did Jody have to pay Tong?

2. Jessica's grandparents gave her $2000 for college to put in a savings account until she starts college in four years. Her grandparents agreed to pay her an additional 7.5% simple interest on the $2000 for every year. How much extra money will her grandparents give her at the end of four years?

3. David read an ad offering $8\frac{3}{4}$ % simple interest on accounts over $500 left for a minimum of 5 years. He has $500 and thinks this sounds like a great deal. How much money will he earn in the 5 years?

4. Javier's parents set an amount of money aside when he was born. They earned 4.5% simple interest on that money each year. When Javier was 15, the account had a total of $1012.50 interest paid on it. How much did Javier's parents set aside when he was born?

5. Kristina received $125 for her birthday. Her parents offered to pay her 3.5% simple interest per year if she would save it for at least one year. How much interest could Kristina earn in one year?

6. Kristina decided she would do better if she put her money in the bank for one year, which paid 2.8% interest compounded annually. Was she right?

7. Suppose Jessica (from problem 2) had put her $2000 in the bank at 3.25% interest compounded annually. How much money would she have earned there at the end of 4 years?

8. Mai put $4250 in the bank at 4.4% interest compounded annually. How much was in her account after 7 years?

9. What is the difference in the amount of money in the bank after five years if $2500 is invested at 3.2% interest compounded annually or at 2.9% interest compounded annually?

10. Ronna was listening to her parents talking about what a good deal compounded interest was for a retirement account. She wondered how much money she would have if she invested $2000 at age 20 at 2.8% interest compounded quarterly and left it until she reached age 65. Determine what the value of the $2000 would become.

Answers

1. $I = 50(0.05)1 = \$2.50$; Jody paid back $52.50.

2. $I = 2000(0.075)4 = \$600$ 3. $I = \$500(0.0875)5 = \$218.75$

4. $\$1012.50 = x(0.045)15$; $x = \$1500$ 5. $I = 125(0.035)1 = \$4.38$

6. $A = 125(1 + 0.028)^1 = \$128.50$, $I = \$3.50$; no, for one year she needs to take the higher interest rate if the compounding is done annually Only after one year will annual compounding earn more than simple interest.

7. $A = 2000(1 + 0.0325)^4 = \$2272.95$, amount earned = $2,275.95 - 2,000 = \$275.95$

8. $A = 4250(1 + 0.044)^7 = \$5745.03$

9. $A = 2500(1 + 0.032)^5 - 2500(1 + 0.029)^5 = \$2926.43 - \$2884.14 = \$42.29$

10. $A = 2000(1 + 0.028)^{180}$ (because $45 \cdot 4 = 180$ quarters) $= \$288,264.15$

EQUIVALENT FRACTIONS

Fractions that name the same value are called **equivalent fractions**, such as $\frac{2}{3} = \frac{6}{9}$.

Three methods for finding equivalent fractions are using a ratio table, the Identity Property of Multiplication (the Giant **1**), and a rectangular area model. The ratio table method is discussed in this guide in the "Ratio" section.

For additional information, see Year 1, Chapter 5, problems GO-14 on page 149 and GO-80 on page 166 or Year 2, Chapter 3, problems MD-27 and 28 on pages 92-93.

THE IDENTITY PROPERTY OF MULTIPLICATION or THE GIANT 1

Multiplying by 1 does not change the value of a number. The Giant **1** uses a fraction that has the same numerator and denominator, such as $\frac{2}{2}$, to find an equivalent fraction.

Example 1

Find three equivalent fractions for $\frac{1}{2}$.

$\frac{1}{2} \cdot \boxed{\frac{2}{2}} = \frac{2}{4}$ $\qquad$ $\frac{1}{2} \cdot \boxed{\frac{3}{3}} = \frac{3}{6}$ $\qquad$ $\frac{1}{2} \cdot \boxed{\frac{4}{4}} = \frac{4}{8}$

Example 2

Use the Giant **1** to find an equivalent fraction to $\frac{7}{12}$ using 96ths: $\frac{7}{12} \cdot \boxed{1} = \frac{?}{96}$

Which Giant **1** do you use?

Since $96 \div 12 = 8$, the Giant **1** is $\frac{8}{8}$: $\qquad$ $\frac{7}{12} \cdot \boxed{\frac{8}{8}} = \frac{56}{96}$

Problems

Use the Giant **1** to find the specified equivalent fraction. Your answer should include the Giant **1** you use and the equivalent numerator.

1. $\frac{4}{3} \cdot \boxed{1} = \frac{?}{15}$ $\qquad$ 2. $\frac{5}{9} \cdot \boxed{1} = \frac{?}{36}$ $\qquad$ 3. $\frac{9}{2} \cdot \boxed{1} = \frac{?}{38}$

4. $\frac{3}{7} \cdot \boxed{1} = \frac{?}{28}$ $\qquad$ 5. $\frac{5}{3} \cdot \boxed{1} = \frac{?}{18}$ $\qquad$ 6. $\frac{6}{5} \cdot \boxed{1} = \frac{?}{15}$

Answers

1. $\frac{5}{5}$, 20   2. $\frac{4}{4}$, 20   3. $\frac{19}{19}$, 171   4. $\frac{4}{4}$, 12   5. $\frac{6}{6}$, 30   6. $\frac{3}{3}$, 18

RECTANGULAR AREA MODEL

The rectangular area model for finding equivalent fractions is based on the area of a rectangle. Draw and shade a rectangle to represent the original fraction. Next, add horizontal lines to the rectangle to divide the area equally so that the rectangle has the same number of equal pieces as the number in the denominator of the second fraction. Note that each rectangle has the same amount of shaded area. Renaming the shaded area in terms of the new, smaller pieces gives the equivalent fraction.

For additional information, see Year 1, Chapter 5, problems GO-56 on page 160 and GO-66 on page 163 or Year 2, Chapter 3, problems MD-40 and 41 on pages 96-97.

Example 1

Use the rectangular area model to find three equivalent fractions for $\frac{1}{2}$.

$\frac{1}{2}$ $\frac{2}{4}$ $\frac{3}{6}$ $\frac{4}{8}$

The number of horizontal rows in the rectangle model corresponds to a Giant **1**, $\frac{2}{2}$, $\frac{3}{3}$, and $\frac{4}{4}$.

Example 2

Use the rectangular area model to find the specified equivalent fraction.

$$\frac{3}{4} \cdot \mathbb{1} = \frac{?}{20}$$

$$\frac{3}{4} \cdot \frac{5}{5} = \frac{15}{20}$$

After drawing the fraction $\frac{3}{4}$, the diagram is divided into five horizontal rows because $20 \div 4$ equals 5. The diagram now shows 15 shaded parts out of 20 total parts. This area model shows the equivalent fractions: $\frac{3}{4} = \frac{15}{20}$

Problems

Draw rectangular models to find the specified equivalent fraction.

1. $\frac{6}{7} = \frac{?}{14}$ 2. $\frac{5}{8} = \frac{?}{32}$ 3. $\frac{2}{9} = \frac{?}{18}$ 4. $\frac{4}{3} = \frac{?}{9}$

Answers

1. $\dfrac{12}{14}$

2. $\dfrac{20}{32}$

3. $\dfrac{4}{18}$

4. $\dfrac{12}{9}$

The following table summarizes the three methods for finding equivalent fractions.

Fraction	Ratio Table	Giant 1	Rectangular Model
$\dfrac{4}{7}$	$\begin{array}{c\|c\|c} 4 & 8 & 12 \\ \hline 7 & 14 & 21 \end{array}$	$\dfrac{4}{7} \cdot \boxed{\dfrac{2}{2}} = \dfrac{8}{14}$ $\dfrac{4}{7} \cdot \boxed{\dfrac{3}{3}} = \dfrac{12}{21}$	$\dfrac{4}{7} = \dfrac{8}{14}$ $\dfrac{4}{7} = \dfrac{12}{21}$

ADDITION AND SUBTRACTION OF FRACTIONS

Before fractions can be added or subtracted, the fractions must have the same denominator, that is, a common denominator. There are three methods for adding or subtracting fractions used in this course.

AREA MODEL METHOD

For additional information, see Year 1, Chapter 7, problems GH-3 through 5 on pages 219-20 or Year 2, Chapter 3, problem MD-56 on page 103.

Step 1: Copy the problem.

$$\frac{1}{3} + \frac{1}{2}$$

Step 2: Draw and divide equal-sized rectangles for each fraction. One rectangle is cut horizontally. The other is cut vertically. Label each rectangle, with the fraction it represents.

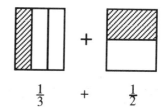

Step 3: Superimpose the lines from each rectangle onto the other rectangle, as if one rectangle is placed on top of the other one.

Step 4:
Rename the fractions as sixths, because the new rectangles are divided into six equal parts. Change the numerators to match the number of sixths in each figure.

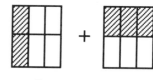

Step 5:
Draw an empty rectangle with sixths, then combine all sixths by shading the same number of sixths in the new rectangle as the total that were shaded in both rectangles from the previous step.

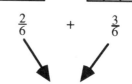

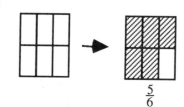

Example 1

$\frac{1}{2} + \frac{2}{5}$ can be modeled as:

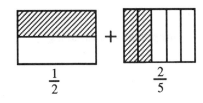

 $+$ ▨ $\quad \frac{5}{10} + \frac{4}{10} \qquad$ so $\qquad$ ▨ $\Rightarrow$ ▨

$\qquad\qquad \frac{5}{10} \qquad\qquad \frac{4}{10} \qquad\qquad\qquad\qquad\qquad\qquad\qquad \frac{9}{10}$

Thus, $\frac{1}{2} + \frac{2}{5} = \frac{9}{10}$.

Example 2

$\frac{1}{2} + \frac{4}{5}$ would be:

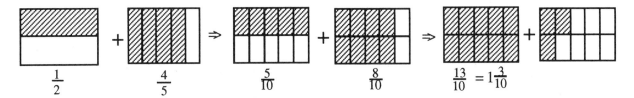

$\qquad \frac{1}{2} \qquad\qquad\qquad \frac{4}{5} \qquad\qquad\qquad \frac{5}{10} \qquad\qquad\qquad \frac{8}{10} \qquad\qquad\qquad \frac{13}{10} = 1\frac{3}{10}$

Problems

Use the area model method to add the following fractions.

1. $\frac{3}{4} + \frac{1}{5}$ 　　　　　　 2. $\frac{1}{3} + \frac{2}{7}$ 　　　　　　 3. $\frac{2}{3} + \frac{3}{4}$

Answers

1. $\frac{19}{20}$ 　　 2. $\frac{13}{21}$ 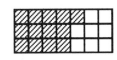 　　 3. $\frac{17}{12} = 1\frac{5}{12}$

IDENTITY PROPERTY OF MULTIPLICATION (Giant **1**) METHOD

The Giant **1**, known in mathematics as the Identity Property of Multiplication, uses a fraction with the same numerator and denominator ($\frac{3}{3}$, for example) to write an equivalent fraction that helps to create common denominators. (Refer to the ratio and fractions sections of this guide for more information.) For additional information, see Year 1, Chapter 7, problems GH-19, 30, 31, and 33 on pages 224-30 or Year 2, Chapter 3, problem MD-68 on page 108.

Example

Add $\frac{1}{3} + \frac{1}{4}$ using the Giant **1**.

Step 1:   Multiply both $\frac{1}{3}$ and $\frac{1}{4}$ by Giant **1**s to get a common denominator.

$$\frac{1}{3} \cdot \frac{4}{4} + \frac{1}{4} \cdot \frac{3}{3} = \frac{4}{12} + \frac{3}{12}$$

Step 2:   Add the numerators of both fractions to get the answer.

$$\frac{4}{12} + \frac{3}{12} = \frac{7}{12}$$

RATIO TABLE METHOD

In Year 1, the least common multiple is found by using ratio tables. The least common multiple is used as the common denominator of the fractions. The Giant **1** or another ratio table can be used to find the new numerators. For additional information, see Year 1, Chapter 6, problem MB-22 on page 187.

Example

Solve $\frac{3}{4} - \frac{2}{7}$ using a ratio table to find the least common denominator of the fractions.

Use a ratio table to find the least common denominator of the fractions. (This is the same as finding the least common multiple of the denominators, 4 and 7.)

4	8	12	16	20	24	(28)
7	14	21	(28)	35	42	49

You then use the Giant **1** to find the new numerator.

$$\frac{3}{4} - \frac{2}{7} \Rightarrow \frac{3}{4} \cdot \frac{7}{7} - \frac{2}{7} \cdot \frac{4}{4} \Rightarrow \frac{21}{28} - \frac{8}{28} \Rightarrow \frac{13}{28}$$

Problems

Find each sum or difference. Use the method of your choice.

1. $\frac{1}{3} + \frac{2}{5}$ 2. $\frac{1}{6} + \frac{2}{3}$ 3. $\frac{3}{8} + \frac{2}{5}$ 4. $\frac{1}{4} + \frac{3}{7}$

5. $\frac{2}{9} + \frac{3}{4}$ 6. $\frac{5}{12} + \frac{1}{3}$ 7. $\frac{4}{5} - \frac{1}{3}$ 8. $\frac{3}{4} - \frac{1}{5}$

9. $\frac{7}{9} - \frac{2}{3}$ 10. $\frac{3}{4} + \frac{1}{3}$ 11. $\frac{5}{6} + \frac{2}{3}$ 12. $\frac{7}{8} + \frac{1}{4}$

13. $\frac{6}{7} - \frac{2}{3}$ 14. $\frac{1}{4} - \frac{1}{3}$ 15. $\frac{3}{5} + \frac{3}{4}$ 16. $\frac{5}{7} - \frac{3}{4}$

17. $\frac{1}{3} - \frac{3}{4}$ 18. $\frac{2}{5} + \frac{9}{15}$ 19. $\frac{3}{5} - \frac{2}{3}$ 20. $\frac{5}{6} - \frac{11}{12}$

Answers

1. $\frac{11}{15}$ 2. $\frac{5}{6}$ 3. $\frac{31}{40}$ 4. $\frac{19}{28}$ 5. $\frac{35}{36}$

6. $\frac{3}{4}$ 7. $\frac{7}{15}$ 8. $\frac{11}{20}$ 9. $\frac{1}{9}$ 10. $\frac{13}{12} = 1\frac{1}{12}$

11. $\frac{3}{2} = 1\frac{1}{2}$ 12. $\frac{9}{8} = 1\frac{1}{8}$ 13. $\frac{4}{21}$ 14. $-\frac{1}{12}$ 15. $\frac{27}{20} = 1\frac{7}{20}$

16. $-\frac{1}{28}$ 17. $-\frac{5}{12}$ 18. $\frac{15}{15} = 1$ 19. $-\frac{1}{15}$ 20. $-\frac{1}{12}$

To summarize addition and subtraction of fractions:

1. Rename each fraction with equivalents that have a common denominator.

2. Add or subtract only the numerators, keeping the common denominator.

3. Simplify if possible.

ADDING AND SUBTRACTING MIXED NUMBERS

To add or subtract mixed numbers, change the mixed numbers into fractions greater than one, find a common denominator, then add or subtract. For additional information, see Year 1, Chapter 7, problems GH-30, 31 and 33 on pages 228-30.

Example 1

Find the sum: $3\frac{1}{5} + 1\frac{2}{3}$.

$3\frac{1}{5} = 3 + \frac{1}{5} \cdot \frac{3}{3} = 3\frac{3}{15}$

$1\frac{2}{3} = 1 + \frac{2}{3} \cdot \frac{5}{5} = +1\frac{10}{15}$

$\overline{\qquad\qquad\qquad\quad 4\frac{13}{15}}$

Example 2

Find the difference: $3\frac{1}{5} - 1\frac{2}{3}$.

$3\frac{1}{5} = \frac{16}{5} \cdot \frac{3}{3} = \frac{48}{15}$

$-1\frac{2}{3} \qquad \frac{5}{3} \cdot \frac{5}{5} = \frac{25}{15}$

$\overline{\qquad\qquad\qquad \frac{23}{15} \text{ or } 1\frac{8}{15}}$

Problems

Find each difference.

1. $2\frac{1}{2} - 1\frac{3}{4}$

2. $4 + 5\frac{3}{7}$

3. $4\frac{1}{3} - 3\frac{5}{6}$

4. $1\frac{1}{6} - \frac{3}{4}$

5. $5\frac{2}{5} - 3\frac{2}{3}$

6. $1\frac{2}{3} + 3\frac{1}{6}$

7. $7 - 1\frac{2}{3}$

8. $5\frac{3}{8} - 2\frac{2}{3}$

9. $3\frac{1}{8} + 1\frac{5}{6}$

Answers

1. $\frac{5}{2} - \frac{7}{4} \Rightarrow \frac{10}{4} - \frac{7}{4} \Rightarrow \frac{3}{4}$

2. $\frac{4}{1} + \frac{38}{7} \Rightarrow \frac{28}{7} + \frac{38}{7} \Rightarrow \frac{66}{7}$ or $9\frac{3}{7}$

3. $\frac{13}{3} - \frac{23}{6} \Rightarrow \frac{26}{6} - \frac{23}{6} \Rightarrow \frac{3}{6}$ or $\frac{1}{2}$

4. $\frac{7}{6} - \frac{3}{4} \Rightarrow \frac{14}{12} - \frac{9}{12} \Rightarrow \frac{5}{12}$

5. $\frac{27}{5} - \frac{11}{3} \Rightarrow \frac{81}{15} - \frac{55}{15} \Rightarrow \frac{26}{15}$ or $1\frac{11}{15}$

6. $\frac{5}{3} + \frac{19}{6} \Rightarrow \frac{10}{6} + \frac{19}{6} \Rightarrow \frac{29}{6}$ or $4\frac{5}{6}$

7. $\frac{7}{1} - \frac{5}{3} \Rightarrow \frac{21}{3} - \frac{5}{3} \Rightarrow \frac{16}{3}$ or $5\frac{1}{3}$

8. $\frac{43}{8} - \frac{8}{3} \Rightarrow \frac{129}{24} - \frac{64}{24} \Rightarrow \frac{65}{24}$ or $2\frac{17}{24}$

9. $\frac{25}{8} + \frac{11}{6} \Rightarrow \frac{75}{24} + \frac{44}{24} \Rightarrow \frac{119}{24}$ or $4\frac{23}{24}$

MULTIPLYING AND DIVIDING FRACTIONS

MULTIPLYING FRACTIONS WITH AN AREA MODEL

Multiplication of fractions is taught using a rectangular area model. Lines that divide the rectangle to represent one fraction are drawn vertically, and the correct number of parts are shaded. Then lines that divide the rectangle to represent the second fraction are drawn horizontally and part of the shaded region is darkened to represent the product of the two fractions. For additional information, see Year 1, Chapter 7, problems GH-43, 46, and 47 on pages 234-36 and problem GH-58 on page 240 or Year 2, Chapter 3, problems MD-40, 41, and 46 on pages 96-99.

Example 1

$\frac{1}{2} \cdot \frac{5}{8}$ (that is, $\frac{1}{2}$ of $\frac{5}{8}$)

Step 1: Draw a unit rectangle and divide it into 8 pieces vertically. Lightly shade 5 of those pieces. Label it $\frac{5}{8}$.

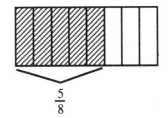

Step 2: Use a horizontal line and divide the unit rectangle in half. Darkly shade $\frac{1}{2}$ of $\frac{5}{8}$ and label it.

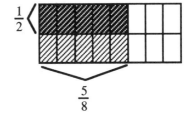

Step 3: Write an equation.

$$\frac{1}{2} \cdot \frac{5}{8} = \frac{5}{16}$$

The rule for multiplying fractions derived from the models above is to multiply the numerators, then multiply the denominators. Simplify the product when possible.

Example 2

a) $\frac{2}{3} \cdot \frac{2}{7} \Rightarrow \frac{2 \cdot 2}{3 \cdot 7} \Rightarrow \frac{4}{21}$

b) $\frac{3}{4} \cdot \frac{6}{7} \Rightarrow \frac{3 \cdot 6}{4 \cdot 7} \Rightarrow \frac{18}{28} \Rightarrow \frac{9}{14}$

Problems

Draw an area model for each of the following multiplication problems and write the answer.

1. $\frac{1}{3} \cdot \frac{1}{6}$

2. $\frac{1}{4} \cdot \frac{3}{5}$

3. $\frac{2}{3} \cdot \frac{5}{9}$

Use the rule for multiplying fractions to find the answer for the following problems. Simplify when possible.

4. $\frac{1}{3} \cdot \frac{2}{5}$

5. $\frac{2}{3} \cdot \frac{2}{7}$

6. $\frac{3}{4} \cdot \frac{1}{5}$

7. $\frac{2}{5} \cdot \frac{2}{3}$

8. $\frac{2}{3} \cdot \frac{1}{4}$

9. $\frac{5}{6} \cdot \frac{2}{3}$

10. $\frac{4}{5} \cdot \frac{3}{4}$

11. $\frac{2}{15} \cdot \frac{1}{2}$

12. $\frac{3}{7} \cdot \frac{1}{2}$

13. $\frac{3}{8} \cdot \frac{4}{5}$

14. $\frac{2}{9} \cdot \frac{3}{5}$

15. $\frac{3}{10} \cdot \frac{5}{7}$

16. $\frac{5}{11} \cdot \frac{6}{7}$

17. $\frac{5}{6} \cdot \frac{3}{10}$

18. $\frac{10}{11} \cdot \frac{3}{5}$

19. $\frac{5}{12} \cdot \frac{3}{5}$

20. $\frac{7}{9} \cdot \frac{5}{14}$

Answers

1. $\frac{1}{18}$

2. $\frac{3}{20}$

3. $\frac{10}{27}$

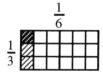

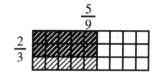

4. $\frac{2}{15}$

5. $\frac{4}{21}$

6. $\frac{3}{20}$

7. $\frac{4}{15}$

8. $\frac{2}{12} = \frac{1}{6}$

9. $\frac{10}{18} = \frac{5}{9}$

10. $\frac{12}{20} = \frac{3}{5}$

11. $\frac{2}{30} = \frac{1}{15}$

12. $\frac{3}{14}$

13. $\frac{12}{40} = \frac{3}{10}$

14. $\frac{6}{45} = \frac{2}{15}$

15. $\frac{15}{70} = \frac{3}{14}$

16. $\frac{30}{77}$

17. $\frac{15}{60} = \frac{1}{4}$

18. $\frac{30}{55} = \frac{6}{11}$

19. $\frac{15}{60} = \frac{1}{4}$

20. $\frac{35}{126} = \frac{5}{18}$

MULTIPLYING MIXED NUMBERS

There are two ways to multiply mixed numbers. One is with generic rectangles. For additional information, see Year 1, Chapter 7, problems GH-60 through 62 on pages 240-41.

Example 1

Find the product: $2\frac{1}{2} \cdot 1\frac{1}{2}$.

Step 1: Draw the generic rectangle. Label the top 1 plus $\frac{1}{2}$. Label the side 2 plus $\frac{1}{2}$.

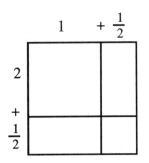

Step 2: Write the area of each smaller rectangle in each of the four parts of the drawing.

Find the total area:

$2 + 1 + \frac{1}{2} + \frac{1}{4} = 3\frac{3}{4}$

Step 3: Write an equation: $2\frac{1}{2} \cdot 1\frac{1}{2} = 3\frac{3}{4}$

Example 2

Find the product: $3\frac{1}{3} \cdot 2\frac{1}{4}$.

$6 + \frac{3}{4} + \frac{2}{3} + \frac{1}{12} \Rightarrow 6 + \frac{9}{12} + \frac{8}{12} + \frac{1}{12} \Rightarrow$
$\Rightarrow 6\frac{18}{12} \Rightarrow 7\frac{1}{2}$

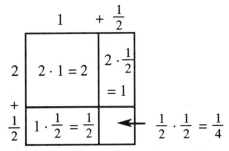

Problems

Use a generic rectangle to find each product.

1. $1\frac{1}{4} \cdot 1\frac{1}{2}$
2. $3\frac{1}{6} \cdot 2\frac{1}{2}$
3. $2\frac{1}{4} \cdot 1\frac{1}{2}$
4. $1\frac{1}{3} \cdot 1\frac{1}{6}$
5. $1\frac{1}{2} \cdot 1\frac{1}{3}$

Answers

1. $1\frac{7}{8}$
2. $7\frac{11}{12}$
3. $3\frac{3}{8}$
4. $1\frac{5}{9}$
5. 2

	1	+	$\frac{1}{2}$
1	1		$\frac{1}{2}$
+ $\frac{1}{4}$	$\frac{1}{4}$		$\frac{1}{8}$

	2	+	$\frac{1}{2}$
3	6		$\frac{3}{2}$
+ $\frac{1}{6}$	$\frac{2}{6}$		$\frac{1}{12}$

	1	+	$\frac{1}{2}$
2	2		1
+ $\frac{1}{4}$	$\frac{1}{4}$		$\frac{1}{8}$

	1	+	$\frac{1}{6}$
1	1		$\frac{1}{6}$
+ $\frac{1}{3}$	$\frac{1}{3}$		$\frac{1}{18}$

	1	+	$\frac{1}{3}$
1	1		$\frac{1}{3}$
+ $\frac{1}{2}$	$\frac{1}{2}$		$\frac{1}{6}$

A second way to multiply mixed numbers is to change them to fractions greater than 1, then multiply the numerators and multiply the denominators. Simplify if possible. For additional information, see Year 1, Chapter 7, problem GH-65 on page 242.

Example 3

$2\frac{1}{2} \cdot 1\frac{1}{4} \Rightarrow \frac{5}{2} \cdot \frac{5}{4} \Rightarrow \frac{5 \cdot 5}{2 \cdot 4} \Rightarrow \frac{25}{8} \Rightarrow 3\frac{1}{8}$

Problems

Find each product, using the method of your choice. Simplify when possible.

1. $2\frac{1}{4} \cdot 1\frac{3}{8}$
2. $3\frac{3}{5} \cdot 2\frac{4}{7}$
3. $2\frac{3}{8} \cdot 1\frac{1}{6}$
4. $3\frac{7}{9} \cdot 2\frac{5}{8}$

5. $1\frac{2}{9} \cdot 2\frac{3}{7}$
6. $3\frac{4}{7} \cdot 5\frac{8}{11}$
7. $2\frac{3}{8} \cdot 1\frac{1}{16}$
8. $2\frac{8}{9} \cdot 2\frac{5}{8}$

9. $1\frac{1}{3} \cdot 1\frac{4}{7}$
10. $2\frac{1}{7} \cdot 2\frac{7}{10}$

Answers

1. $3\frac{3}{32}$
2. $9\frac{9}{35}$
3. $2\frac{37}{48}$
4. $9\frac{11}{12}$
5. $2\frac{61}{63}$

6. $20\frac{5}{11}$
7. $2\frac{67}{128}$
8. $7\frac{7}{12}$
9. $2\frac{2}{21}$
10. $5\frac{11}{14}$

DIVIDING FRACTIONS USING AN AREA MODEL

Division of fractions is introduced with a rectangular area model. The division problem $8 \div 2$ means, "In 8, how many groups of 2 are there?" Similarly, $\frac{1}{2} \div \frac{1}{4}$ means, "In $\frac{1}{2}$, how many fourths are there?" For additional information, see Year 1, Chapter 7, problems GH-89 and 91 on page 249 or Year 2, Chapter 7, problem CT-32 on page 258.

Example 1

Use the rectangular model to divide: $\frac{1}{2} \div \frac{1}{4}$.

Step 1: Using the rectangle, we first divide it into 2 equal pieces. Each piece represents $\frac{1}{2}$. Shade $\frac{1}{2}$ of it.

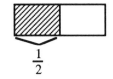

Step 2: Then divide the <u>original</u> rectangle into four equal pieces. Each section represents $\frac{1}{4}$.

In the shaded section, $\frac{1}{2}$, there are 2 fourths.

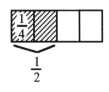

Step 3: Write the equation.

$$\frac{1}{2} \div \frac{1}{4} = 2$$

Example 2

In $\frac{3}{4}$, how many $\frac{1}{2}$s are there?

That is, $\frac{3}{4} \div \frac{1}{2} = ?$

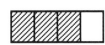 Start with $\frac{3}{4}$.

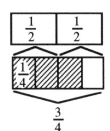

In $\frac{3}{4}$ there is one full $\frac{1}{2}$ shaded and half of another one (that is half of one-half).

So: $\frac{3}{4} \div \frac{1}{2} = 1\frac{1}{2}$
(one and one-half halves)

Problems

Use the rectangular model to divide.

1. $1\frac{1}{3} \div \frac{1}{6}$ 2. $\frac{3}{2} \div \frac{3}{4}$ 3. $1 \div \frac{1}{4}$ 4. $1\frac{1}{4} \div \frac{1}{2}$ 5. $2\frac{2}{3} \div \frac{1}{6}$

Answers

1.   8   2.   2   3.   4   4. $2\frac{1}{2}$ 5.   16

DIVIDING FRACTIONS USING RECIPROCALS

Two numbers that have a product of 1 are reciprocals. For example, $\frac{1}{4} \cdot \frac{4}{1} = 1$, $\frac{1}{2} \cdot \frac{2}{1} = 1$, and $\frac{1}{6} \cdot \frac{6}{1} = 1$, so $\frac{1}{4}$ and $\frac{4}{1}$, $\frac{1}{2}$ and $\frac{2}{1}$, and $\frac{1}{6}$ and $\frac{6}{1}$ are all reciprocals. For additional information, see Year 1, Chapter 7, problem GH-94 on page 250 or Year 2, Chapter 7, problem CT-22 on page 256.

There is another way to divide fractions: invert the divisor, that is, write its reciprocal. (The divisor is the number after the division sign.) After inverting the divisor, change the division sign to a multiplication sign and multiply. Simplify if possible.

Example 1

$\frac{3}{4} \div \frac{1}{2} \Rightarrow \frac{3}{4} \cdot \frac{2}{1} \Rightarrow \frac{6}{4}$ or $1\frac{1}{2}$

Example 2

$1\frac{1}{3} \div \frac{1}{6} \Rightarrow \frac{4}{3} \cdot \frac{6}{1} \Rightarrow \frac{24}{3}$ or 8

The examples above were written horizontally, but division of fractions problems can also be written in the vertical form such as $\dfrac{\frac{1}{2}}{\frac{1}{4}}$, $\dfrac{\frac{3}{4}}{\frac{1}{2}}$, and $\dfrac{1\frac{1}{3}}{\frac{1}{6}}$. They still mean the same thing:

$\dfrac{\frac{1}{2}}{\frac{1}{4}}$ means, "In $\frac{1}{2}$, how many $\frac{1}{4}$s are there?" $\dfrac{\frac{3}{4}}{\frac{1}{2}}$ means, "In $\frac{3}{4}$, how many $\frac{1}{2}$s are there?"

$\dfrac{1\frac{1}{3}}{\frac{1}{6}}$ means, "In $1\frac{1}{3}$, how many $\frac{1}{6}$s are there?"

You can use a Super Giant **1** to solve these vertical division problems. This Super Giant **1** uses the reciprocal of the divisor. For additional information, see Year 1, Chapter 7, problems GH-116 and 120 on pages 256-57 or Year 2, Chapter 7, problems CT-36 on page 259 and problem 48 on page 262.

Example 3

$$\frac{\frac{1}{2}}{\frac{1}{4}} \cdot \frac{\frac{4}{1}}{\frac{4}{1}} = \frac{\frac{4}{2}}{1} = \frac{4}{2} = 2$$

Example 4

$$\frac{\frac{3}{4}}{\frac{1}{2}} \cdot \frac{\frac{2}{1}}{\frac{2}{1}} = \frac{\frac{6}{4}}{1} = \frac{6}{4} = \frac{3}{2} = 1\frac{1}{2}$$

Example 5

$$1\frac{1}{3} = \frac{\frac{4}{3}}{\frac{1}{6}} = \frac{\frac{4}{3}}{\frac{1}{6}} \cdot \frac{\frac{6}{1}}{\frac{6}{1}} = \frac{\frac{24}{3}}{1} = \frac{24}{3} = 8$$

Example 6

$$\frac{2}{3} \div \frac{1}{5} = \frac{2}{3} \cdot \frac{5}{1} = \frac{10}{3} = 3\frac{1}{3}$$

Compared to:

$$\frac{\frac{2}{3}}{\frac{1}{5}} \cdot \frac{\frac{5}{1}}{\frac{5}{1}} = \frac{\frac{10}{3}}{1} = 3\frac{1}{3}$$

Examples 1 and 4 and Examples 2 and 5 result in the same answer, because dividing by a number is the same as multiplying by its reciprocal.

Problems

Solve these division problems. Use any method.

1. $\frac{3}{7} \div \frac{1}{8}$ 2. $1\frac{3}{7} \div \frac{1}{2}$ 3. $\frac{4}{7} \div \frac{1}{3}$ 4. $1\frac{4}{7} \div \frac{1}{3}$ 5. $\frac{6}{7} \div \frac{5}{8}$

6. $\frac{3}{10} \div \frac{5}{7}$ 7. $2\frac{1}{3} \div \frac{5}{8}$ 8. $7 \div \frac{1}{3}$ 9. $1\frac{1}{3} \div \frac{2}{5}$ 10. $2\frac{2}{3} \div \frac{3}{4}$

11. 4 12. 3 13. $\frac{5}{8} \div 1\frac{1}{4}$ 14. $10\frac{1}{3} \div \frac{1}{6}$ 15. $\frac{3}{5} \div 6$

Answers

1. $3\frac{3}{7}$ 2. $2\frac{6}{7}$ 3. $1\frac{5}{7}$ 4. $4\frac{5}{7}$ 5. $1\frac{13}{35}$

6. $\frac{21}{50}$ 7. $3\frac{11}{15}$ 8. 21 9. $3\frac{1}{3}$ 10. $3\frac{5}{9}$

11. $1\frac{9}{11}$ 12. $\frac{1}{3}$ 13. $\frac{1}{2}$ 14. 62 15. $\frac{1}{10}$

FRACTION, DECIMAL, AND PERCENT EQUIVALENTS

Fractions, decimals, and percents are different ways to represent the same number.

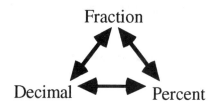

For additional information, see Year 1, Chapter 3, problems PR-25 and 26 on page 75, Chapter 6, problem MB-23 on page 187, problem MB-48 on page 194, problem MB-64 on page 199, and problem MB-83 on page 204 or Year 2, Chapter 3, problem MD-109 on page 119.

Examples

Decimal to percent:

Multiply the decimal by 100.

$(0.34)(100) = 34\%$

Percent to decimal:

Divide the percent by 100.

$78.6\% \div 100 = 0.786$

Fraction to percent:

Write a proportion to find an equivalent fraction using 100 as the denominator. The numerator is the percent.

$$\frac{4}{5} = \frac{x}{100} \text{ so } \frac{4}{5} = \frac{80}{100} = 80\%$$

Percent to fraction:

Use 100 as the denominator. Use the percent as the numerator. Simplify as needed.

$$22\% = \frac{22}{100} = \frac{11}{50}$$

Decimal to fraction:

Use the decimal as the numerator. Use the decimal place value name as the denominator. Simplify as needed.

a) $0.2 = \frac{2}{10} = \frac{1}{5}$ b) $0.17 = \frac{17}{100}$

Fraction to decimal:

Divide the numerator by the denominator.

$$\frac{3}{8} = 3 \div 8 = 0.375$$

Problems

Convert the fraction, decimal, or percent as indicated.

1. Change $\frac{1}{4}$ to a decimal.

2. Change 50% into a fraction in lowest terms.

3. Change 0.75 to a fraction in lowest terms.

4. Change 75% to a decimal.

5. Change 0.38 to a percent.

6. Change $\frac{1}{5}$ to a percent.

7. Change 0.3 to a fraction.

8. Change $\frac{1}{8}$ to a decimal.

9. Change $\frac{1}{3}$ to a decimal.

10. Change 0.08 to a percent.

11. Change 87% to a decimal.

12. Change $\frac{3}{5}$ to a percent.

13. Change 0.4 to a fraction in lowest terms.

14. Change 65% to a fraction in lowest terms.

15. Change $\frac{1}{9}$ to a decimal.

16. Change 125% to a fraction in lowest terms.

17. Change $\frac{8}{5}$ to a decimal.

18. Change 3.25 to a percent.

19. Change $\frac{1}{16}$ to a decimal. Change the decimal to a percent.

20. Change $\frac{1}{7}$ to a decimal.

21. Change 43% to a fraction. Change the fraction to a decimal.

22. Change 0.375 to a percent. Change the percent to a fraction.

23. Change $\frac{7}{8}$ to a decimal. Change the decimal to a percent.

Answers

1. 0.25

2. $\frac{1}{2}$

3. $\frac{3}{4}$

4. 0.75

5. 38%

6. 20%

7. $\frac{3}{10}$

8. 0.125

9. $0.\bar{3}$

10. 8%

11. 0.87

12. 60%

13. $\frac{2}{5}$

14. $\frac{13}{20}$

15. $0.\bar{1}$

16. $1\frac{5}{4}$

17. 1.6

18. 325%

19. 0.0625; 6.25%

20. ≈ 0.143

21. $\frac{43}{100}$; 0.43

22. 37.5%; $\frac{3}{8}$

23. 0.875; 87.5%

In the expression 5^2, 5 is the **base** and 2 is the **exponent**. For x^a, x is the base and a is the exponent. 5^2 means $5 \cdot 5$ and 5^3 means $5 \cdot 5 \cdot 5$.

$\frac{5^5}{5^2}$ (which can also be written $5^5 \div 5^2$) means $\frac{5 \cdot 5 \cdot 5 \cdot 5 \cdot 5}{5 \cdot 5}$.

Use the Giant **1** to find the numbers in common. There are two Giant **1**s, namely, $\frac{5}{5}$ twice so $\frac{5 \cdot 5 \cdot 5 \cdot 5 \cdot 5}{5 \cdot 5} = 5^3$ or 125. Writing 5^3 is usually sufficient.

When there is a variable, it is treated the same way.

$\frac{x^7}{x^3}$ means $\frac{x \cdot x \cdot x \cdot x \cdot x \cdot x \cdot x}{x \cdot x \cdot x}$. The Giant **1** here is $\frac{x}{x}$ (three of them). The answers is x^4.

$5^2 \cdot 5^3$ means $(5 \cdot 5)(5 \cdot 5 \cdot 5)$ which is 5^5.

$(5^2)^3$ means $(5^2)(5^2)(5^2)$ or $(5 \cdot 5)(5 \cdot 5)(5 \cdot 5)$ which is 5^6.

When the problems have variables such as $x^4 \cdot x^5$, you only need to add the exponents. The answer is x^9. If the problem is $(x^4)^5$ (x^4 to the fifth power) it means $x^4 \cdot x^4 \cdot x^4 \cdot x^4 \cdot x^4$. The answer is x^{20}. You multiply exponents in this case.

If the problem is $\frac{x^{10}}{x^4}$, you subtract the bottom exponent from the top exponent $(10 - 4)$.

The answer is x^6. You can also have problems like $\frac{x^{10}}{x^{-4}}$. You still subtract: $10 - (-4)$ is 14, and the answer is x^{14}.

You need to be sure the bases are the same to use these laws. $x^5 \cdot y^6$ cannot be simplified.

In general the laws of exponents are:

$$x^a \cdot x^b = x^{(a+b)} \qquad (x^a)^b = x^{ab} \qquad \frac{x^a}{x^b} = x^{(a-b)}$$

$$x^0 = 1 \qquad (x^a y^b)^c = x^{ac} y^{bc}$$

These rules are true if $x \neq 0$ and $y \neq 0$.

For additional information, see Year 2, Chapter 8, problem GS-26 on page 298 and Chapter 10, problem MG-35 on page 377.

Examples

a) $x^8 \cdot x^7 = x^{15}$

b) $\dfrac{x^{19}}{x^{13}} = x^6$

c) $(z^8)^3 = z^{24}$

d) $(x^2y^3)^4 = x^8y^{12}$

e) $\dfrac{x^4}{x^{-3}} = x^7$

f) $(2x^2y^3)^2 = 4x^4y^6$

g) $(3x^2y^{-2})^3 = 27x^6y^{-6}$ or $\dfrac{27x^6}{y^6}$

h) $\dfrac{x^8y^5z^2}{x^3y^6z^{-2}} = \dfrac{x^5z^4}{y}$ or $x^5y^{-1}z^4$

Problems

Simplify each expression.

1. $5^2 \cdot 5^4$

2. $x^3 \cdot x^4$

3. $\dfrac{5^{16}}{5^{14}}$

4. $\dfrac{x^{10}}{x^6}$

5. $(5^3)^3$

6. $(x^4)^3$

7. $(4x^2\,y^3)^4$

8. $\dfrac{5^2}{5^{-3}}$

9. $5^5 \cdot 5^{-2}$

10. $(y^2)^{-3}$

11. $(4a^2b^{-2})^3$

12. $\dfrac{x^5y^4z^2}{x^4y^3z^2}$

13. $\dfrac{x^6y^2z^3}{x^{-2}y^3z^{-1}}$

14. $4x^2 \cdot 2x^3$

Answers

1. 5^6

2. x^7

3. 5^2

4. x^4

5. 5^9

6. x^{12}

7. $256x^8y^{12}$

8. 5^5

9. 5^3

10. y^{-6} or $\dfrac{1}{y^6}$

11. $64a^6b^{-6}$ or $\dfrac{64a^6}{b^6}$

12. xy

13. $\dfrac{x^8z^4}{y}$ or $x^8y^{-1}z^4$

14. $8x^5$

SCIENTIFIC NOTATION

Scientific notation is a way of writing very large and very small numbers compactly. A number is said to be in scientific notation when it is written as the product of two factors as described below.

- The first factor is less than 10 and greater than or equal to 1.
- The second factor has a base of 10 and an integer exponent (power of 10).
- The factors are separated by a multiplication sign.
- A positive exponent indicates a number whose absolute value is greater than one.
- A negative exponent indicates a number whose absolute value is less than one.

Scientific Notation	Standard Form
$5.32 \cdot 10^{11}$	532,000,000,000
$2.61 \cdot 10^{-15}$	0.00000000000000261

It is important to note that the exponent does not necessarily mean to use that number of zeros.

The number $5.32 \cdot 10^{11}$ means $5.32 \cdot 100,000,000,000$. Thus, two of the 11 places in the standard form of the number are the 3 and the 2 in 5.32. Standard form in this case is 532,000,000,000. In this example you are moving the decimal point to the right 11 places to find standard form.

The number $2.61 \cdot 10^{-15}$ means $2.61 \cdot 0.000000000000001$.
Moving the decimal point to the left 15 places to find standard form.
Here the standard form is 0.00000000000000261.

For additional information, see Year 2, Chapter 10, problem MG-65 on page 386.

Example 1

Write each number in standard form.

$7.84 \cdot 10^8 \implies 784,000,000$ and $3.72 \cdot 10^{-3} \implies 0.00372$

When taking a number in standard form and writing it in scientific notation, remember there is only <u>one</u> digit before the decimal point, that is, the number must be between 1 and 9, inclusive.

Example 2 $52,050,000 \implies 5.205 \cdot 10^7$ and $0.000372 \implies 3.72 \cdot 10^{-4}$

The exponent denotes the number of places to move the decimal point in the standard form. In the first example above, the decimal point is at the end of the number and it was moved 7 places. In the second example above, the exponent is negative because the original number is very small, that is, less than one.

Problems

Write each number in standard form.

1. $7.85 \cdot 10^{11}$ 2. $1.235 \cdot 10^9$ 3. $1.2305 \cdot 10^3$ 4. $3.89 \cdot 10^{-7}$ 5. $5.28 \cdot 10^{-4}$

Write each number in scientific notation.

6. 391,000,000,000   7. 0.0000842   8. 123,056.7   9. 0.000000502

10. 25.7   11. 0.035   12. 5,600,000   13. 1346.8

14. 0.000000000006   15. 634,700,000,000,000

Note:
On a scientific or graphing calculator, displays like 4.357^{12} and 3.65^{-03} are numbers expressed in scientific notation. The first number means $4.357 \cdot 10^{12}$ and the second means $3.65 \cdot 10^{-3}$. The calculator docs this because there is not enough room on its display window to show the entire number.

Answers

1. 785,000,000,000   2. 1,235,000,000   3. 1230.5

4. 0.000000389   5. 0.000528   6. $3.91 \cdot 10^{11}$

7. $8.42 \cdot 10^{-5}$ 8. $1.230567 \cdot 10^5$ 9. $5.02 \cdot 10^{-7}$

10. $2.57 \cdot 10^1$ 11. $3.5 \cdot 10^{-2}$ 12. $5.6 \cdot 10^6$

13. $1.3468 \cdot 10^3$ 14. $6.0 \cdot 10^{-12}$ 15. $6.347 \cdot 10^{14}$

In every Diamond Problem, the product of the two side numbers (left and right) is the top number and their sum is the bottom number.

Diamond Problems are an excellent way of practicing addition, subtraction, multiplication, and division of both positive and negative integers, decimals and fractions. They have the added benefit of preparing students for factoring binomials in algebra.

For more information, see Year 1, Chapter 5, problem GO-12 on page 148 or Year 2, Chapter 1, problem GO-63 on page 25.

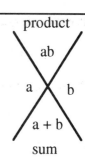

Example 1

The top number is the product of -20 and 10, or -200. The bottom number is the sum of -20 and 10, or -20 + 10 = -10.

Example 2

The product of the right number and -2 is 8. Thus, if you divide 8 by - 2 you get - 4, the right number.
The sum of -2 and - 4 is - 6, the bottom number.

Example 3

To get the left number, subtract 4 from 6, 6 – 4 = 2. The product of 2 and 4 is 8, the top number.

Example 4

The easiest way to find the side numbers in a situation like this one is to look at all the pairs of factors of - 8. They are:

- 1 and 8, -2 and 4, -4 and 2, and -8 and 1.

Only one of these pairs has a sum of 2: - 2 and 4. Thus, the side numbers are - 2 and 4.

Problems

Complete each of the following diamond problems.

1. 　　2. 　　3. 　　4.

5. 　　6. 　　7. 　　8.

9. 　　10. 　　11. 　　12.

13. 　　14. 　　15. 　　16.

Answers

1. -32 and -4　　　　2. -4 and -6　　　　3. -6 and 6　　　　4. 6 and -1

5. 4.56 and 5　　　　6. 5 and 40.5　　　　7. 3.4 and 11.56　　8. 3 and 6.2

9. $-\frac{1}{14}$ and $-\frac{5}{14}$　10. $\frac{13}{10}$ and $\frac{13}{50}$　11. 2 and $\frac{29}{10}$ or $2\frac{9}{10}$　12. $\frac{1}{3}$ and $\frac{1}{3}$

13. xy and x + y　　14. a and 2a　　　15. -6b and $-48b^2$　16. 4a and $12a^2$

Algebra

and

Functions

Order of Operations

Combining Like Terms

Rectangular Model of Multiplication & the Distributive Property

Solving Word Problems Using Guess and Check Tables

Solving Equations

Graphing

Linear Functions

Inequalities

For problems like $3 + 4 \cdot 2$, some students think the answer is 14 and some think the answer is 11. There needs to be a method to simplify an expression which involves more than one operation so that everyone can agree on the answer.

There is a set of rules to follow that provides a consistent way for everyone to evaluate expressions. These rules, called the **order of operations,** must be followed in order to arrive at a correct solution. As the name says, these rules tell the order in which the mathematical operations are done.

For additional information, see Year 1, Chapter 3, problems PR-38 through 43 and PR-45 on pages 78-81 or Year 2, Chapter 4, problems GC-2, 3, 5, and 6 on pages 129-31.

The first step is to organize the expression into parts called TERMS. Terms are separated by addition (+) or subtraction (–) symbols <u>unless</u> the addition (+) or subtraction (–) happens inside parentheses.

Examples of numerical terms are: 4, $3(6)$, $6(9 - 4)$, $2 \cdot 3^2$, $3(5 + 2^3)$, and $\frac{16 - 4}{6}$.

For the problem above, $3 + 4 \cdot 2$, the terms are circled at right. $\boxed{3} + \boxed{4 \cdot 2}$

Each term is simplified separately, giving $3 + 8$. Then the terms are added: $3 + 8 = 11$. Thus, $3 + 4 \cdot 2 = 11$.

Example 1

$$2 \cdot 3^2 + 3(6 - 3) + 10$$

• Circle the terms.

$$\boxed{2 \cdot 3^2} + \boxed{3(6 - 3)} + 10$$

• Simplify each term until it is one number.

  • The operations inside the parentheses are done first.

$$\boxed{2 \cdot 3^2} + \boxed{3(3)} + 10$$

  • Exponents are a form of multiplication.

$$\boxed{2 \cdot 9} + \boxed{3(3)} + 10$$

  • Multiplication and division are done from left to right.

$$\boxed{18} + \boxed{9} + 10$$

• Finally, add or subtract the terms going from left to right.

$$27 + 10$$
$$37$$

Example 2

$$5 - 8 \div 2^2 + 6(5 + 4) - 5^2$$

- Circle the terms.

$$\boxed{5} - \boxed{8 \div 2^2} + \boxed{6(5 + 4)} - \boxed{5^2}$$

- Simplify inside the parentheses.

$$\boxed{5} - \boxed{8 \div 2^2} + \boxed{6(9)} - \boxed{5^2}$$

- Simplify the exponents.

$$\boxed{5} - \boxed{8 \div 4} + \boxed{6(9)} - 25$$

- Multiply and divide from left to right.

$$\boxed{5} - \boxed{2} + \boxed{54} - \boxed{25}$$

- Finally, add and subtract from left to right

$$3 + 54 - 25$$
$$57 - 25$$
$$32$$

Example 3

$$20 + \frac{5 + 7}{3} - 4^2 + 12 \div 4$$

- Circle the terms.

$$\boxed{20} + \boxed{\frac{5 + 7}{3}} - \boxed{4^2} + \boxed{12 \div 4}$$

- Multiply and divide left to right, including exponents. (Note: calculation details are shown for the second term.)

$$\boxed{20} + \boxed{\frac{5 + 7}{3} = \frac{12}{3} = 4} - \boxed{16} + \boxed{3}$$

$$24 - 16 + 3$$

- Add or subtract from left to right.

$$8 + 3$$
$$11$$

Problems

Circle the terms, then simplify the expression.

1. $5 \cdot 3 + 4$

2. $10 \div 5 + 3$

3. $2(9 - 4) \cdot 7$

4. $6(7 + 3) + 8 \div 2$

5. $15 \div 3 + 7(8 + 1) - 6$

6. $\frac{9}{3} + 5 \cdot 3^2 - 2(14 - 5)$

7. $\frac{20}{6 + 4} + 7 \cdot 2 \div 2$

8. $\frac{5 + 30}{7} + 6^2 - 18 \div 9$

9. $2^3 + 8 - 16 \div 8 \cdot 2$

10. $25 - 5^2 + 9 - 3^2$

11. $5(17 - 7) + 4 \cdot 3 - 8$

12. $(5 - 2)^2 + (9 + 1)^2$

13. $4^2 + 9(2) \div 6 + (6 - 1)^2$

14. $\frac{18}{3^2} + \frac{5 \cdot 3}{5}$

15. $3(7 - 2)^2 + 8 \div 4 - 6 \cdot 5$

16. $14 \div 2 + 6 \cdot 8 \div 2 - (9 - 3)^2$

17. $\frac{27}{3} + 18 - 9 \div 3 - (3 + 4)^2$

18. $26 \cdot 2 \div 4 - (6 + 4)^2 + 3(5 - 2)^3$

19. $\left(\frac{42 + 3}{5}\right)^2 + 3^2 - (5 \cdot 2)^2$

Answers

1. 19	2. 5	3. 70	4. 64	5. 62
6. 30	7. 9	8. 39	9. 12	10. 0
11. 54	12. 109	13. 44	14. 5	15. 47
16. -5	17. -25	18. -6	19. -10	

COMBINING LIKE TERMS

The order of operations rules are used to simplify or evaluate a numerical expression. Algebraic expressions can also be simplified by combining (adding or subtracting) terms that have the same variable(s) into one quantity. Note that "same variable" means that the variable <u>and</u> its exponent are the same. The skill of combining like terms is necessary for solving equations. For additional information, see Year 1, Chapter 8, problems MC-70 through 73 on pages 297-98 or Year 2, Chapter 4, problems GC-42 and 43 on pages 140-41.

Example 1

Combine like terms to simplify the expression $3x + 5x + 7x$.

All these terms have x as the variable, so they are combined into one term, $15x$.

Example 2

Simplify the expression $3x + 12 + 7x + 5$.

The terms with x can be combined. The terms without variables (the constants) can also be combined.

$3x + 12 + 7x + 5$
$3x + 7x + 12 + 5$ Note that in the simplified form the term with the variable is listed before
$10x + 17$ the constant term.

Example 3

Simplify the expression $5x + 4x^2 + 10 + 2x^2 + 2x - 6 + x - 1$.

$5x + 4x^2 + 10 + 2x^2 + 2x - 6 + x - 1$ Note that terms with different exponents are not
$4x^2 + 2x^2 + 5x + 2x + x + 10 - 6 - 1$ combined and are listed in order of decreasing
$6x^2 + 8x + 3$ power of the variable, in simplified form, with
     the constant term last.

Example 4

In *Foundations for Algebra: Year 2,* algebra tiles are used to model how to combine like terms.

The large square ▢ represents x^2.

The rectangle ▯ represents x.

The small square ▫ represents one.

We can only combine tiles that are alike:

▢ with ▢ , ▯ with ▯ , and ▫ with ▫

If we use the two sets of algebra tiles inside the dashed line boundaries, shown at right and below, and write an algebraic expression for each of them, we get $2x^2 + 3x + 4$ and $3x^2 + 5x + 7$. Use the tiles to help combine the like terms:

$2x^2$ (2 large squares) + 3x (3 rectangles) + 4 (4 small squares)
+ $3x^2$ (3 large squares) + 5x (5 rectangles) + 7 (7 small squares)

The combination of the two sets of tiles, written algebraically, is:

$$5x^2 + 8x + 11$$

Example 5

Sometimes it is helpful to take an expression that is written horizontally, circle the terms with their signs, and rewrite <u>like</u> terms in vertical columns before you combine them:

$$(2x^2 - 5x + 6) + (3x^2 + 4x - 9)$$

(2x²) (−5x) (+6) + (3x²) (+4x) (−9)

$$\begin{array}{rrr} 2x^2 & -5x & +6 \\ +3x^2 & +4x & -9 \\ \hline 5x^2 & -x & -3 \end{array}$$

This procedure may make it easier to identify the terms as well as the sign of each term.

Problems

Combine the following sets of terms.

1. $(2x^2 + 6x + 10) + (4x^2 + 2x + 3)$

2. $(3x^2 + x + 4) + (x^2 + 4x + 7)$

3. $(8x^2 + 3) + (4x^2 + 5x + 4)$

4. $(4x^2 + 6x + 5) - (3x^2 + 2x + 4)$

5. $(4x^2 - 7x + 3) + (2x^2 - 2x - 5)$

6. $(3x^2 - 7x) - (x^2 + 3x - 9)$

7. $(5x + 6) + (-5x^2 + 6x - 2)$

8. $2x^2 + 3x + x^2 + 4x - 3x^2 + 2$

9. $3c^2 + 4c + 7x - 12 + (-4c^2) + 9 - 6x$

10. $2a^2 + 3a^3 - 4a^2 + 6a + 12 - 4a + 2$

Answers

1. $6x^2 + 8x + 13$ 2. $4x^2 + 5x + 11$ 3. $12x^2 + 5x + 7$ 4. $x^2 + 4x + 1$

5. $6x^2 - 9x - 2$ 6. $2x^2 - 10x + 9$ 7. $-5x^2 + 11x + 4$ 8. $7x + 2$

9. $-c^2 + 4c + x - 3$ 10. $3a^3 - 2a^2 + 2a + 14$

RECTANGULAR MODEL FOR MULTIPLICATION AND THE DISTRIBUTIVE PROPERTY

BASE TEN MODEL

In Year 1, Chapter 4 the rectangular model for is developed to give students a geometric view of multiplication. This approach will lead to multiplying polynomials in algebra. Base ten blocks are used in some classes to help students build a rectangular model of multiplication.

A set of base ten blocks looks like the example shown below:

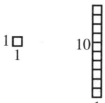

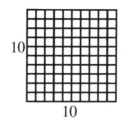

cube = 1   "long" = 10 cubes   "flat" = 10 "longs" or 100 cubes
                                          or a combination of "longs" and "unit" cubes

Example 1

To show the multiplication problem 4 · 17, base ten blocks are used to build the model below.

Build 4 · 10.   (four base ten longs = 4 · 10)

Build 4 · 7.   (28 base ten cubes = 4 · 7)

Put the two sets of blocks
together to show a model
of 4 · (10 + 7) or 4 · 17.

The dimensions of the rectangle are
4 · (10 + 7) or 4 · 17.

The area of the rectangle is 40 + 28 = 68 square units.

Example 2

The model below shows that 12 · 13 is equal to (10 + 2) · (10 + 3).

The rectangular regions at right
are the subproblems (parts of
problems) for the multiplication
problem.

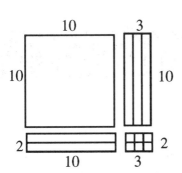

subproblems

$$10 \cdot 10 = 100$$
$$10 \cdot 3 = 30$$
$$2 \cdot 10 = 20$$
$$2 \cdot 3 = +6$$
156 square units

If the areas of all the rectangular regions are added together, the total area will
be 156 square units.

When the traditional multiplication algorithm is used, the product looks like this:

```
  13
 x12
  26
 130
 156
```

Example 3

The next example shows the generic rectangular models of 6 · 32 and 15 · 25.

6 · 32 = 6 · (30 + 2)   15 · 25 = (10 + 5) · (20 + 5)

	30	+ 2
6	180	12

= 180 + 12 = 192

	20	+ 5
10	200	50
+ 5	100	25

200 + 50 + 100 + 25 = 375

Problems

Solve the following multiplication problems by drawing generic rectangles.

1. $2 \cdot 24$ 2. $3 \cdot 92$ 3 $5 \cdot 89$ 4. $11 \cdot 23$

5. $25 \cdot 32$ 6. $17 \cdot 58$ 7. $27 \cdot 65$ 8. $32 \cdot 32$

9. $18 \cdot 57$ 10. $11 \cdot 321$ 11. $23 \cdot 252$ 12. $35 \cdot 411$

13. $115 \cdot 227$ 14. $20 \cdot 115$ 15. $101 \cdot 310$ 16. $215 \cdot 205$

17. $107 \cdot 35$ 18. $255 \cdot 810$ 19. $909 \cdot 780$

Answers

1. 48

	20	+ 4
2	40	8

2. 276

	90	+ 2
3	270	6

3. 445

	80	+ 9
5	400	45

4. 253

	20	+ 3
10	200	30
+ 1	20	3

5. 800

	30	+ 2
20	600	40
+ 5	150	10

6. 986

	50	+ 8
10	500	80
+ 7	350	56

7. 1755

	60	+ 5
20	1200	100
+ 7	420	35

8. 1024

	30	+ 2
30	900	60
+ 2	60	4

9. 1026

	50	+ 7
10	500	70
+ 8	400	56

10. 3531

	300	+ 20	+ 1
10	3000	200	10
+ 1	300	20	1

11. 5796

	200	+ 50	+ 2
20	4000	1000	40
+ 3	600	150	6

12. 14,385

	400	+ 10	+ 1
30	12000	300	30
+ 5	2000	50	5

13. 26,105

	200	+ 20	+ 7
100	20000	2000	700
+ 10	2000	200	70
+ 5	1000	100	35

14. 2300

	100	+ 10	+ 5
20	2000	200	100

15. 31,310

	300	+ 10
100	30000	1000
+ 1	300	10

16. 44,075

	200	+ 5
200	40000	1000
+ 10	2000	50
+ 5	1000	25

17. 3745

	30	+ 5
100	3000	500
+ 7	210	35

18. 206,550

	800	+ 10
200	160,000	2000
+ 50	40000	500
+ 5	4000	50

19. 709,020

	700	+ 80
900	630,000	72,000
+ 9	6300	720

DISTRIBUTIVE PROPERTY WITH INTEGERS

The Distributive Property may be used to multiply numbers by breaking them into smaller parts. Once the smaller numbers are multiplied, the products are then added or subtracted to get the final answer.

- Rewrite the multiplication problem as two or more terms of products to be added or subtracted. Examples: $4 \cdot 28 = 4 \cdot 20 + 4 \cdot 8$ and $6 \cdot 372 = 6 \cdot 300 + 6 \cdot 70 + 6 \cdot 2$
- Multiply the separate parts. Examples: $80 + 32$ and $1800 + 420 + 12$
- Add or subtract the products. Examples: $4 \cdot 28 = 112$ and $6 \cdot 372 = 2232$

For additional information, see Year 2, Chapter 4, problem GC-18 on page 134.

>>Examples follow on the next page.>>

Example 1

$6 \cdot 37$

30	+	7

6 {
30		7
30		7
30		7
30		7
30		7
30		7

$6 \cdot 37$

$= 60(30 + 7)$

$= 6 \cdot 30 + 6 \cdot 7$

$= 180 + 42$

$= 222$

$6 \cdot 37 = 222$

Example 2

$3 \cdot 715$

700	+	10	+	5

3 {
700		10		5
700		10		5
700		10		5

$3 \cdot 715$

$= 3(700 + 10 + 5)$

$= 3(700) + 3(10) + 3(5)$

$= 2100 + 30 + 15$

$= 2145$

$3 \cdot 715 = 2145$

Example 3

$8 \cdot 22$

20	+	2

8 {
20		2
20		2
20		2
20		2
20		2
20		2
20		2
20		2

$8 \cdot 22$

$= 8(20 + 2)$

$= 8(20) + 8(2)$

$= 160 + 16$

$= 176$

$8 \cdot 22 = 176$

Example 4

$6 \cdot 98$

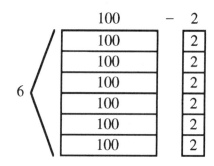

100	−	2

6 {
100		2
100		2
100		2
100		2
100		2
100		2

$6 \cdot 98$

$= \ 6(100 - 2)$

$= \ 6(100) - 6(2)$

$= \ 600 - 12$

$= \ 588$

$6 \cdot 98 = 588$

Algebra and Functions

Problems

Find each product using one of the methods shown in the examples.

1. 2 · 93

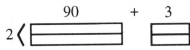

2. 7 · 36

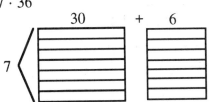

3. 5 · 199

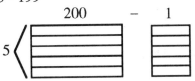

4. 4 · 398

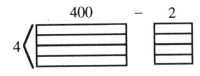

5. 6 · 536

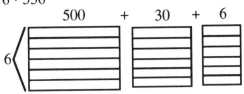

6. 3 · 328

7. 2 · 597

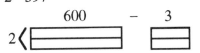

8. 5 · 43

9. 6 · 91
10. 8 · 214
11. 6 · 537
12. 3 · 193

13. 50 · 139
14. 40 · 27
15. 20 · 38
16. -3 · 81

17. -8 · 273
18. 15 · 21
19. 12 · 39
20. 3 · 495

Answers

1. 2(90 + 3)
   2(90) + 2(3)
   180 + 6 = 186

   2 · 93 = 186

2. 7(30 + 6)
   7(30) + 7(6)
   210 + 42 = 252

   7 · 36 = 252

3. 5(200 − 1)
   5(200) − 5(1)
   1000 − 5 = 995

   5 · 199 = 995

4. 4(400 − 2)
   4(400) −4(2)
   1600 − 8 = 1592

   4 · 398 = 1592

5.   6(500 + 30 + 6)   6 · 536 = 3216   6.   3(300 + 20 + 8)   3 · 328 = 984
      6(500) + 6(30) + 6(6)   3(300) +3(20) + 3(8)
      3000 + 180 + 36 = 3216   900 + 60 + 24 = 984

7.   2(600 –3)   2 · 597 = 1194
      2(600) – 2(3)
      1200 – 6 = 1194

8.   5(40 + 3) = 5(40) + 5(3) = 200 + 15 = 215

9.   6(90 + 1) = 6(90) + 6(1) = 540 + 6 = 546

10.   8(214) = 8(200 + 10 + 4) = 8(200) + 8(10) + 8(4) = 1600 + 80 + 32 = 1712

11.   6(500 + 30 + 7) = 6(500) + 6(30) + 6(7) = 3000 + 180 + 42 = 3222

12.   3(100 + 90 + 3) = 3(100) + 3(90) + 3(3) = 300 + 270 + 9 = 579

13.   50(100 + 30 + 9) = 50(100) + 50(30) + 50(9) = 5000 + 1500 + 450 = 6950

14.   40(20 + 7) = 40(20) + 40(7) = 800 + 280 = 1080

15.   20(30 + 8) = 20(30) + 20(8) = 600 + 160 = 760

16.   -3(80 + 1) = -3(80) + -3(1) = -240 + -3 = -243

17.   -8(200) + -8(70) + -8(3) = -1600 + -560 + -24 = -2184

18.   15(20 + 1) = 15(20) + 15(1) = 300 + 15 = 315

19.   12(30 + 9) = 12(30) + 12(9) = 360 + 108 = 468

20.   3(500 – 5) = 3(500) – 3(5) = 1500 – 15 = 1485

DISTRIBUTIVE PROPERTY WITH ALGEBRA TILES

The Distributive Property can be used to multiply expressions containing variables. For example, 5(x + 3) means there are 5 groups. Each group contains one "x" and 3 "ones." If each group contains one "x" and 3 "ones," then the 5 groups would contain a total of 5 "x" and 15 "ones." Using the Distributive Property, 5(x + 3) can be rewritten as 5(x) + 5(3), which equals 5x + 15. We can use algebra tiles to model this idea.

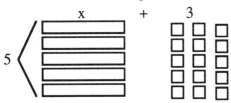

If you count the total number of tiles, there are 5 x's and 15 "ones."

$$5(x + 3) = 5(x) + 5(3) = 5x + 15$$

The area of the rectangular arrangement of algebra tiles can be represented three different ways.

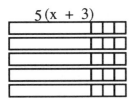 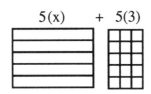 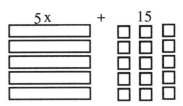

For additional information, see Year 2, Chapter 4, problems GC-56 and 57 on page 145.

Example 1

Model the three representations of the expression 3(x + 7) using algebra tiles. Remember that 3(x + 7) means there are three groups, and each group contains one x and 7 ones. If you count the number of each kind of tile used in each of the representations, there are 3 x's and 21 ones. Therefore, we can say 3(x + 7) = 3(x) + 3(7) = 3x + 21.

3(x + 7)

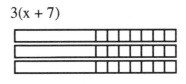

= 3(x) + 3(7)

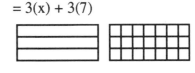

= 3x + 21

Example 2

Model the three representations of the expression $5(x + 4)$. Remember that $5(x + 4)$ means there are five groups, and each group contains one x and 4 ones. You will use 5 x's and 20 ones each time. Therefore we can say $5(x + 4) = 5(x) + 5(4) = 5x + 20$.

$5(x + 4)$ $= 5(x) + 5(4)$ $= 5x + 20$

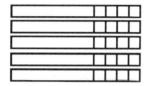

Problems

Show the distributed form of each expression below.

1. $2(x + 5)$ 2. $4(x + 4)$ 3. $3(x + 2)$ 4. $5(x + 3)$ 5. $9(x + 1)$

Answers

1. $2(x) + 2(5) = 2x + 10$ 2. $4(x) + 4(4) = 4x + 16$ 3. $3(x) + 3(2) = 3x + 6$

4. $5(x) + 5(3) = 5x + 15$ 5. 5. $9(x) + 9(1) = 9x + 9$

Example 3

Model the three representations of the expression $2(3x + 2)$ using algebra tiles. This time there are two groups, and each group contains 3 x's and 2 ones. You will use a total of 6 rectangular strips and 4 ones for each representation. Therefore we can say $2(3x + 2) = 2(3x) + 2(2) = 6x + 4$.

$2(3x + 2)$

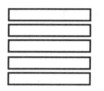

$= 2(3x) + 2(2)$

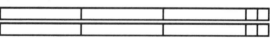

$= 6x + 4$

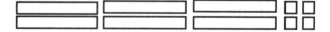

Problems

Show the distributed form of each expression below.

1. $2(2x + 4)$ 2. $5(3x + 1)$ 3. $3(4x + 2)$ 4. $7(2x + 1)$

5. $4(5x + 2)$ 6. $2(6x + 5)$ 7. $3(3x + 6)$

Answers

1. $2(2x) + 2(4) = 4x + 8$

2. $5(3x) + 5(1) = 15x + 5$

3. $3(4x) + 3(2) = 12x + 6$

4. $7(2x) + 7(1) = 14x + 7$

5. $4(5x) + 4(2) = 20x + 8$

6. $2(6x) + 2(5) = 12x + 10$

7. $3(3x) + 3(6) = 9x + 18$

Example 4

Model the three representations of the expression $3(3x - 4)$ using algebra tiles. Remember that $3(3x - 4)$ means there are three groups and each group contains 3 x's and 4 negative ones. You will use a total of 9 x's and 12 negative ones for each representation. Therefore we can say that $3(3x - 4) = 3(3x) + 3(-4) = 9x - 12$. Notice that the negative ones are shaded.

$3(3x - 4)$

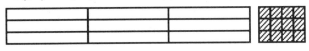

$= 3(3x) + 3(-4)$

$= 9x - 12$

Problems

Show the distributed form of each expression below, that is, simplify each expression.

1. $2(x - 4)$

2. $5(2x - 1)$

3. $6(3x - 3)$

4. $4(x - 9)$

5. $3(5x - 3)$

6. $10(4x - 7)$

7. $2(x^2 + 4x - 3)$

8. $3(x^2 - 2x + 5)$

Answers

1. $2(x) + 2(-4) = 2x - 8$

2. $5(2x) + 5(-1) = 10x - 5$

3. $6(3x) + 6(-3) = 18x - 18$

4. $4(x) + 4(-9) = 4x - 36$

5. $3(5x) + 3(-3) = 15x - 9$

6. $10(4x) + 10(-7) = 40x - 70$

7. $2(x^2) + 2(4x) + 2(-3) = 2x^2 + 8x - 6$

8. $3(x^2) + 3(-2x) + 3(5) = 3x^2 - 6x + 15$

FACTORING

The Distributive Property is used to simplify expressions in Year 2, Chapter 4. Students need to be able to take the simplified form and reverse the order to put the expressions into the factored form. The factors show what was multiplied before the distribution took place.

For additional information, see Year 2, Chapter 4, problems GC-31 on page 137 and GC-109 on page 160.

Example 1

$3x + 3$ is the simplified (distributed) form of a product. $3(x + 1)$ is the factored form.
In other words, since $3x + 3 = 3 \cdot x + 3 \cdot 1$, the 3 is placed outside the parentheses: $3(x + 1)$.

Example 2

The simplified expression for $6x + 12$ can be written $2(3x + 6)$ or $3(2x + 4)$ or $6(x + 2)$. This example has many possible answers because 6 and 12 have many common factors. In general, an expression is completely factored when the greatest common factor is used. Thus, the factored form of $6x + 12$ is $6(x + 2)$.

Problems

Factor.

1. $3x + 18$	2. $5x + 15$	3. $6b + 24$	4. $3x - 6$	5. $6a + 3$
6. $5x + 60$	7. $12x + 48$	8. $7x - 21$	9. $-3x - 9$	10. $-4x - 8$

Answers

1. $3(x + 6)$	2. $5(x + 3)$	3. $6(b + 4)$	4. $3(x - 2)$
5. $3(2a + 1)$	6. $5(x + 12)$	7. $12(x + 4)$	8. $7(x - 3)$
9. $3(-x - 3)$ or $-3(x + 3)$		10. $-4(x + 2)$ or $4(-x - 2)$	

GUESS AND CHECK

WHY WE USE GUESS AND CHECK TABLES

Guess and Check is one method that students can use to solve many types of problems, especially word problems. When students create a Guess and Check table, it provides a record of the student's thinking. The patterns in the table lead directly to writing algebraic equations for the word problems.

Writing equations is one of the most important algebra skills students learn. Guess and Check tables help make this skill accessible to all students. In order to help students see the relationships in a word problem, we require them to include at least four entries (rows) in their Guess and Check tables. The repetition of the operations is needed to see how the columns are related. After students have several weeks of practice using Guess and Check tables to solve problems, we begin generalizing from the patterns in the table of guesses to write an equation that represents the relationships in the problem. During the first few weeks, we intentionally use problems that might be solved with fewer than four guesses so that students can focus on constructing effective tables and later using the patterns to write equations. Thus early in the course, if the correct answer is found earlier than the fourth entry, a number of incorrect entries should be written to complete the table.

We also believe that writing the answer in a sentence after the table is complete is important because many students forget what the question actually asked. The sentence helps the student see the "big picture" and brings closure to the problem.

In Year 1, refer to Chapter 3, problem PR-2 on pages 68-69, for a detailed, step-by-step demonstration of a Guess and Check problem similar to the one below and to problem PR-5 on page 70 for the Tool Kit entry for Guess and Check tables.

In Year 2, refer to Chapter 1, problem GO-17 on pages 9-10, for a detailed, step-by-step demonstration of a Guess and Check problem similar to the one below and to problem GO-33 on page 15 for the Tool Kit for Guess and Check tables.

Example 1

A box of fruit has three times as many nectarines as grapefruit. Together there are 36 pieces of fruit. How many pieces of each type of fruit are there?

Step 1: Guess the number of fruit you know the least about.

Guess Number of Grapefruit	
11	

Step 2: What else do we need to know?

The number of nectarines, which is three times the number of grapefruit.

Guess Number of Grapefruit	Number of Nectarines	
11	3(11) = 33	

Step 3:   What else do we need to know?

The total number of pieces of fruit.

Guess Number of Grapefruit	Number of Nectarines	Total Pieces of Fruit	
11	33	11 + 33 = 44	

Step 4:   What else do we need to know?

We need to check the total pieces of fruit based on a guess of 11 grapefruit and compare it to the total given in the problem.

Guess Number of Grapefruit	Number of Nectarines	Total Pieces of Fruit	Check 36
11	33	44	too high

Step 5:   Determine our next guess. Our total was 44; the total needed is 36, so our guess was too high and our next guess should be lower.

Guess Number of Grapefruit	Number of Nectarines	Total Pieces of Fruit	Check 36
11	33	44	too high
10	30	40	too high

Step 6:   Determine our next guess. Our total was 40; the total needed is 36, so our guess was too high and our next guess should be still lower.

Guess Number of Grapefruit	Number of Nectarines	Total Pieces of Fruit	Check 36
11	33	44	too high
10	30	40	too high
8	24	32	too low

Step 7:   Determine our next guess. Our total was 32; the total needed is 36, so our guess was too low and our next guess should be higher than 8 but lower than 10.

Guess Number of Grapefruit	Number of Nectarines	Total Pieces of Fruit	Check 36
11	33	44	too high
10	30	40	too high
8	24	32	too low
9	27	36	correct

There are 9 grapefruit and 27 nectarines in the box.

Example 2

The perimeter of a rectangle is 120 feet. If the length of the rectangle is ten feet more than the width, what are the dimensions (length and width) of the rectangle?

Step 1: Guess the width because, of the two required answers, it is the one we know the least about. The length is 10 feet more than the width, so add 10 to the guess.

Guess Width	Length	Perimeter	Check 120
10	20	$(10 + 20) \cdot 2 = 60$	too low

Step 2: Since the guess of 10 resulted in an answer that is too low, we should increase the guess. Paying close attention to each guess and its result helps determine the next guess to narrow down the possible guesses to reach the answer. Note: as students get more experience using Guess and Check tables, they learn to make better educated guesses from one step to the next to solve problems quickly or to establish the pattern they need to write an equation.

Guess Width	Length	Perimeter	Check 120
10	20	$(10 + 20) \cdot 2 = 60$	too low
20	30	100	too low
30	40	140	too high
25	35	120	correct

The dimensions are 25 and 35 feet.

Example 3

Jorge has some dimes and quarters. He has 10 more dimes than quarters and the collection of coins is worth $2.40. How many dimes and quarters does Jorge have?

NOTE: This type of problem is more difficult than others because the number of things asked for is different than their value. Separate columns for each part of the problem must be added to the table as shown below. Students often neglect to write the third and fourth columns.

Guess Number of Quarters	Number of Dimes	Value of Quarters	Value of Dimes	Total Value	Check $2.40
10	20	2.50	2.00	4.50	too high
8	18	2.00	1.80	3.80	too high
6	16	1.50	1.60	3.10	too high
4	14	1.00	1.40	2.40	correct

Jorge has four quarters and 14 dimes.

71

HELPFUL QUESTIONS TO ASK YOUR CHILD

When your child is having difficulty with a Guess and Check problem, it may be because he/she does not understand the problem, not because he/she does not understand the Guess and Check process. Here are some helpful questions to ask when your child does not understand the problem. (These are useful in non-Guess and Check situations, too.)

1. What are you being asked to find?

2. What information have you been given?

3. Is there any unneeded information? If so, what is it?

4. Is there any necessary information that is missing? If so, what information do you need?

TIPS ABOUT COLUMN TITLES

1. Guess the answer to the question. Ten or the student's age are adequate first guesses.

2. Continue establishing columns by asking, "What else do we need to know to determine whether our guess is correct or too low or too high?"

3. After the guess, only put the answer to one calculation in each column. Students sometimes try to put the answer to several mental calculations in one column. (See NOTE in Example 3 on the previous page.)

Problems

Solve these problems using Guess and Check tables. Write each answer in a sentence.

1. A wood board 100 centimeters long is cut into two pieces. One piece is 26 centimeters longer than the other. What are the lengths of the two pieces?

2. Thu is five years older than her brother Tuan. The sum of their ages is 51. What are their ages?

3. Tomas is thinking of a number. If he triples his number and subtracts 13, the result is 305. Of what number is Tomas thinking?

4. Two consecutive numbers have a sum of 123. What are the two numbers?

5. Two consecutive even numbers have a sum of 246. What are the numbers?

6. Joe's age is three times Aaron's age and Aaron is six years older than Christina. If the sum of their ages is 99, what is Christina's age? Joe's age? Aaron's age?

7. Farmer Fran has 38 barnyard animals, consisting of only chickens and goats. If these animals have 116 legs, how many of each type of animal are there?

8. A wood board 156 centimeters long is cut into three parts. The two longer parts are the same length and are 15 centimeters longer than the shortest part. How long are the three parts?

9. Juan has 15 coins, all nickels and dimes. This collection of coins is worth 90¢. How many nickels and dimes are there? (Hint: Create separate column titles for, "Number of Nickels," "Value of Nickels," "Number of Dimes," and "Value of Dimes.")

10. Tickets to the school play are $ 5.00 for adults and $ 3.50 for students. If the total value of all the tickets sold was $2517.50 and 100 more students bought tickets than adults, how many adults and students bought tickets?

11. A wood board 250 centimeters long is cut into five pieces: three short ones of equal length and two that are both 15 centimeters longer than the shorter ones. What are the lengths of the boards?

12. Conrad has a collection of three types of coins: nickels, dimes, and quarters. There is an equal amount of nickels and quarters but three times as many dimes. If the entire collection is worth $ 9.60, how many nickels, dimes, and quarters are there?

Answers

1. The lengths of the boards are 37 cm and 63 cm.

2. Thu is 28 years old and her brother is 23 years old.

3. Tomas is thinking of the number 106.

4. The two consecutive numbers are 61 and 62.

5. The two consecutive numbers are 122 and 124.

6. Christina is 15, Aaron is 21, and Joe is 63 years old.

7. Farmer Fran has 20 goats and 18 chickens.

8. The lengths of the boards are 42, 57, and 57 cm.

9. Juan has 12 nickels and 3 dimes.

10. There were 255 adult and 355 student tickets purchased for the play.

11. The lengths of the boards are 44 and 59 cm.

12. Conrad has 16 nickels, 16 quarters and 48 dimes.

WRITING EQUATIONS FROM A GUESS AND CHECK TABLE

Guess and Check is a powerful strategy for solving problems. In fact, some types of problems in Algebra 2 are easier to solve this way than by writing an equation. However, solving complicated problems with a Guess and Check table can be time consuming and it may be difficult to find the correct solution if it is not an integer. The patterns developed in the table can be generalized by using variables to write equations. Writing and solving the equation is often more efficient than guessing and checking.

Additional examples are found in the Year 2 text, Chapter 4, problem GC-83 on page 152. Consider two of the examples from the previous Guess and Check discussion.

Example 1

Guess Number of Grapefruit	Number of Nectarines	Total Pieces of Fruit	Check 36
11	33	44	too high
10	30	40	too high

After several guesses and checks establish a pattern in the problem, you can generalize it using a variable. Since we could guess any number of grapefruit, use x to represent it. The pattern for the number of nectarines is three times the number of grapefruit or $3x$. The total pieces of fruit is the sum of column one and column two, so our table becomes:

Guess Number of Grapefruit	Number of Nectarines	Total Pieces of Fruit	Check 36
x	3x	x + 3x	= 36

Since we want the total to agree with the check, our equation is $x + 3x = 36$. Simplifying this yields $4x = 36$, so $x = 9$ (grapefruit) and then $3x = 27$ (nectarines).

The pattern and structure of the Guess and Check table allowed us to write and solve an equation, so guessing was no longer necessary.

Example 2

Guess Width	Length	Perimeter	Check 120
10	20	$(10 + 20) \cdot 2 = 60$	too low
20	30	100	too low

Again, since we could guess any width, we labeled this column x. The pattern for the second column is that it is 10 more than the first: $x + 10$. Perimeter is found by multiplying the sum of the width and length by 2. Our table now becomes:

Guess Width	Length	Perimeter	Check 120
x	x + 10	$(x + x + 10) \cdot 2$	= 120

Solving the equation:

$$(x + x + 10) \cdot 2 = 120$$
$$2x + 2x + 20 = 120$$
$$4x + 20 = 120$$
$$4x = 100 \quad \text{So} \quad x = 25 \text{ (width) and } x + 10 = 35 \text{ (length)}$$

Algebra and Functions

Problems

Write an equation for each Guess and Check table from the previous Problems section, numbers 1-12.

Answers (may vary)

1. $x + (x + 26) = 100$

2. $x + (x + 5) = 51$

3. $3x - 13 = 305$

4. $x + (x + 1) = 123$

5. $x + (x + 2) = 246$

6. $x + (x + 6) + 3(x + 6) = 149$

7. $2x + 4(38 - x) = 116$

8. $x + (x + 15) + (x + 15) = 156$

9. $0.05x + 0.10(15 - x) = 0.90$

10. $\$5x + \$3.50(x + 100) = 2517.50$

11. $3x + 2(x + 15) = 250$

12. $0.05x + 0.25x + 0.10(3x) = 9.60$

SOLVING EQUATIONS

BALANCE SCALE METHOD

An equation is a mathematical sentence with an equal sign.

One aid to understanding solving equations is using the idea of a balance scale. Sometimes equations have unknowns on both sides. The idea is to keep each side of the scale even or equal. No matter what equation is given, it is treated like a balance scale. Negative and positive signs or tiles are used to signify negative and positive numbers. Cups (unknowns) are used to hide the amount of tiles used on either side of the equation (balance scale).

Additional information is in Year 2, Chapter 5, problems MC-38 and 39 on page 181 and problems MC-56 through 58 on page 185.

Example

Solve the following equation using the balance scale method. (Remember that the cups are each hiding the same amount of tiles.) An unknown must be the same number within each equation.

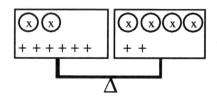

$2x + 6 = 4x + 2$

This equation represents the balance scale at left.

We can remove two cups from both sides of the scale (–2x).

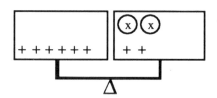

Subtract 2x from both sides.

$6 = 2x + 2$

We can remove two tiles from each side (–2).

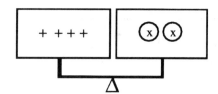

Subtract 2 from both sides.

$4 = 2x$

Divide both sides by 2: $\frac{4}{2}$ and $\frac{2x}{2}$.

If 4 tiles equal 2 cups, there must be 2 tiles in each cup.

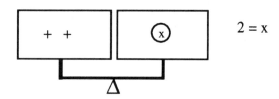

$2 = x$

Since you have found $x = 2$, put 2 back in the original equation and check.

$$2x + 6 = 4x + 2$$
$$2(2) + 6 = 4(2) + 2$$
$$4 + 6 = 8 + 2$$
$$10 = 10 \ \sqrt{}$$

Problems

Solve these equations. Remember to check your answers.

1. $4x + 3 = 2x + 7$ 2. $4x - 3 = 7x + 9$ 3. $5b - 7 = 9b + 13$

4. $12c + 3 = 17c - 2$ 5. $9x - 4 = 6x - 7$ 6. $13n + 12 = 19n - 6$

7. $13n - 12 = 6n + 9$ 8. $4c + 16 = 3c + 15$ 9. $-5c - 16 = 2c + 5$

10. $10x + 18 = 8x - 12$

Answers

1. $x = 2$ 2. $x = -4$ 3. $b = -5$ 4. $c = 1$ 5. $x = -1$

6. $n = 3$ 7. $n = 3$ 8. $c = -1$ 9. $c = -3$ 10. $x = -15$

COVER-UP METHOD

Another method for solving simple equations is the <u>cover-up</u> method. Students cover up the unknown part of the equation and ask what would have to be uncovered to make the equation true.

For additional information, see Year 1, Chapter 2, problem GS-74 on pages 55-56 or Year 2, Chapter 5, problems MC-14 and 15 on pages 175-76.

Example 1

Use the cover-up method to solve the one-step equation below.

$x + 5 = 17$
$(?) + 5 = 17$ Cover up the x.
$x = 12$ What number plus 5 equals 17?
$(12) + 5 = 17$ Put 12 where the x was.

Example 2

Use the cover-up method to solve the two-step equation below.

$2x + 4 = 10$
$(?) + 4 = 10$ Cover the term with x in it.
$(6) + 4 = 10$ What number plus 4 equals 10?
$2x = 6$ 6 plus 4 equals 10, so 2x must equal 6.
$2 \cdot (?) = 6$ 2 times what number equals 6?
$2 \cdot (3) = 6$ 2 times 3 equals 6, so
$x = 3$ x must equal 3.

Problems

Solve these equations. Remember to check your answers.

1. $2x + 3 = 15$ 2. $10n - 13 = 37$ 3. $14 - 7n = 42$

4. $72 = 4x + 4$ 5. $3 = 2x - 11$ 6. $402n - 412 = 2000$

7. $25y + 6 = 181$ 8. $24 = 2x + 10$ 9. $-42y + 24 = -102$

10. $11m + 321 = 409$

Answers

1. $x = 6$ 2. $n = 5$ 3. $n = -4$ 4. $x = 17$ 5. $x = 7$

6. $n = 6$ 7. $y = 7$ 8. $x = 7$ 9. $y = 3$ 10. $m = 8$

INVERSE OPERATIONS

The algebraic approach to solving equations uses inverse operations to "undo" what has been done to create the equation. By systematically working backward, the value of the variable can be found. Multiplication and division are inverse operations, as are addition and subtraction.

The order of operations in the equation $5x + 7 = 37$ is to multiply by 5 and then add 7 to get 37. To solve the equation, work backward using inverse operations. First subtract 7 to get 30, then divide by 5 to get 6.

The examples below model the steps that solve equations algebraically. For additional information, see Year 1, Chapter 2, problem GS-75 on page 56.

Example 1

5x means 5 times some number x. For example, suppose $5x = 20$. To solve, we do the inverse of multiplying by 5, which is dividing by 5. So $20 \div 5 = 4$, $x = 4$.

Example 2

$3x = 30$ Divide by 3 to solve. The result is $x = 10$.

$\frac{x}{3} = 30$ To solve, do the inverse of dividing by 3; that is, multiply by 3. The result is $x = 90$.

Example 3

$x + 3 = 30$ Solve by subtracting 3: $x = 27$.

Example 4

$x - 3 = 30$ Solve by adding 3: $x = 33$.

Example 5

$3x + 9 = 15$ Subtract 9.
$3x = 6$ Divide by 3.
$x = 2$
$3(2) + 9 = 15$ Substitute 2 for x to check your answer.

Example 6

$\frac{n}{3} + 6 = 10$ Subtract 6.

$\frac{n}{3} = 4$ Multiply by 3.

$n = 12$

$\frac{(12)}{3} + 6 = 10$ Substitute to check.

Problems

Solve these equations. Remember to check your answers.

1. $2x + 3 = 7$
2. $5x - 3 = 7$
3. $4x + 6 = 10$
4. $3x - 6 = 6$
5. $x - 10 = 25$
6. $x - 10 = -5$
7. $\frac{x}{2} + 3 = 8$
8. $\frac{n}{3} - 6 = 9$
9. $201n + 602 = 1808$
10. $10x + 15 = 165$
11. $12x + 12 = 144$
12. $23x - 9 = 106$
13. $401c - 201 = 3007$
14. $43 = 4y - 5$
15. $16 + 9y = 88$
16. $32 + 6y = 74$
17. $301 + 302c = 1207$
18. $55n - 52 = 553$
19. $\frac{x}{5} - 6 = -9$
20. $\frac{x}{20} - 7 = -2$

Answers

1. $x = 2$
2. $x = 2$
3. $x = 1$
4. $x = 4$
5. $x = 35$
6. $x = 5$
7. $x = 10$
8. $n = 45$
9. $n = 6$
10. $x = 15$
11. $x = 11$
12. $x = 5$
13. $c = 8$
14. $y = 12$
15. $y = 8$
16. $y = 7$
17. $c = 3$
18. $n = 11$
19. $x = -15$
20. $x = 100$

DISTRIBUTIVE PROPERTY

You can also solve equations that involve the Distributive Property.

Example 1

Solve the equation $3(5x + 2) = 8x + 20$.

$$3(5x + 2) = 8x + 20$$
$$3 \cdot 5x + 3 \cdot 2 = 8x + 20$$

Use the Distributive Property to eliminate the parentheses.

$$15x + 6 = 8x + 20$$
$$ - 6 - 6$$

Subtract 6 from both sides of the equation.

$$15x = 8x + 14$$
$$-8x -8x$$

Subtract 8x from both sides of the equation.

$$\frac{7x}{7} = \frac{14}{7}$$

Divide both sides by 7 to get one x.

$$x = 2$$

Substitute the solution into the original equation to check your answer.

$$3(5 \cdot 2 + 2) = 8 \cdot 2 + 20$$

Use the order of operations and simplify.

$$3(12) = 36 \qquad \sqrt{}$$

Example 2

$$2(x + 3) = 4(x + 1)$$
$$2 \cdot x + 2 \cdot 3 = 4 \cdot x + 4 \cdot 1$$

$$2x + 6 = 4x + 4$$
$$-4 -4$$

$$2x + 2 = 4x$$
$$-2x -2x$$

$$\frac{2}{2} = \frac{2x}{2}$$

$$1 = x$$

Check:

$$2(1 + 3) = 4(1 + 1)$$

$$2(4) = 4(2)$$

$$8 = 8 \quad \sqrt{}$$

Problems

Solve these equations. Remember to check your answers.

1. $3(2x - 5) = 21$ 2. $2(4x + 5) = 26$ 3. $7(2x - 4) = 42$

4. $3(2x + 2) = 12$ 5. $5(10x - 6) = 300$ 6. $-4(2x - 6) = 24$

7. $2(3x - 2) = 4(x + 2)$ 8. $-5(3x + 2) = -3(x - 2)$ 9. $4(3x - 2) = -12$

10. $-4(-3x - 2) = 12$

Answers

1. $x = 6$ 2. $x = 2$ 3. $x = 5$ 4. $x = 1$ 5. $x = 6.6$

6. $x = 0$ 7. $x = 6$ 8. $x = -1\frac{1}{3}$ 9. $x = -\frac{1}{3}$ 10. $x = \frac{1}{3}$

RATIO EQUATIONS (PROPORTIONS)

To solve proportions, use cross multiplication to remove the fractions and then solve in the usual way. Cross multiplication may <u>only</u> be used to solve proportions; that is, cross multiplication may only be used when each side of the equation is a ratio. Multiply the numerator of the left ratio by the denominator of the right ratio, write an equal sign, then write the product of the other numerator and denominator.

For additional information, see Year 1, Chapter 6, problem MB-8 on page 183 or Year 2, Chapter 6, problem RS-23 on page 213.

Example 1

$$\frac{m}{6} = \frac{15}{9}$$
$$9 \cdot m = 6 \cdot 15$$
$$9m = 90$$
$$m = 10$$

Example 2

$$\frac{6}{\$9.75} = \frac{8.5}{w}$$
$$6w = \$82.875$$
$$w = \$13.81$$

Example 3

$$\frac{x + 1}{3} = \frac{x - 2}{5}$$
$$5(x + 1) = 3(x - 2)$$
$$5x + 5 = 3x - 6$$
$$2x + 5 = -6$$
$$2x = -11$$
$$x = \frac{-11}{2} = -5.5$$

Problems

Solve these equations. Remember to check your answers.

1. $\dfrac{2}{5} = \dfrac{y}{15}$

2. $\dfrac{x}{36} = \dfrac{4}{9}$

3. $\dfrac{2}{300} = \dfrac{6}{m}$

4. $\dfrac{5}{8} = \dfrac{x}{100}$

5. $\dfrac{6}{\$9.75} = \dfrac{9}{x}$

6. $\dfrac{8}{2000} = \dfrac{m}{3000}$

7. $\dfrac{8}{\$13.25} = \dfrac{x}{\$24.95}$

8. $\dfrac{8}{14} = \dfrac{4}{m}$

9. $\dfrac{20}{30} = \dfrac{50}{x}$

10. $\dfrac{40}{100} = \dfrac{45}{x}$

11. $\dfrac{x-1}{4} = \dfrac{7}{8}$

12. $\dfrac{3y}{5} = \dfrac{24}{10}$

13. $\dfrac{x}{x+1} = \dfrac{3}{5}$

14. $\dfrac{3}{y} = \dfrac{6}{y-2}$

15. $\dfrac{1}{x} = \dfrac{5}{x+1}$

Answers

1. $y = 6$

2. $x = 16$

3. $m = 900$

4. $x = 62.5$

5. $x \approx \$14.63$

6. $m = 12$

7. $x \approx 15.06$

8. $m = 7$

9. $x = 75$

10. $x = 112.5$

11. $x = 4.5$

12. $y = 4$

13. $x = 1\frac{1}{2}$

14. $y = -2$

15. $x = \frac{1}{4}$

COMBINING LIKE TERMS BEFORE SOLVING

Example 1

$$\begin{aligned}
3x + 2x - 8 &= -x - 8 + 12 \\
5x - 8 &= -x + 4 \\
6x - 8 &= 4 \\
6x &= 12 \\
x &= 2
\end{aligned}$$

Example 2

$$\begin{aligned}
-2(x - 3) + 4x &= -(-x + 1) \\
-2x + 6 + 4x &= x - 1 \\
2x + 6 &= x - 1 \\
x + 6 &= -1 \\
x &= -7
\end{aligned}$$

Problems

Solve these equations.

1. $3x + 2x + 2 = -x + 14$

2. $x + 2(x - 1) = x + 10$

3. $6x + 4x - 2 = 15$

4. $6x - 3x + 2 = -10$

5. $x + 8 + x - 6 = 3(x - 5)$

6. $3(m - 2) = -2(m - 7)$

7. $4 - 6(w + 2) = 10$

8. $6 - 2(x - 3) = 12$

Answers

1. $x = 2$
2. $x = 6$
3. $x = 1.7$
4. $x = -4$

5. $x = 17$
6. $m = 4$
7. $w = -3$
8. $x = 0$

EQUATIONS WITH MORE THAN ONE VARIABLE (FORMULAS OR LITERAL EQUATIONS)

Follow the same procedures for solving multi-variable equations as shown in the preceding sections of the guide.

For additional information, see Year 2, Chapter 7, problems CT-112 through 114 on page 279.

Example 1

$x + 3 = 5$ $x + b = 5$

subtract 3   subtract b
$x = 5 - 3 = 2$ $x = 5 - b$ (done)

Example 2

$2y = 17$ $dy = 17$

divide by 2   divide by d
$y = \frac{17}{2} = 8\frac{1}{2}$ $y = \frac{17}{d}$ (done)

Example 3

$2x - 1 = 5$ $2x - b = 7$
$2x = 6$ $2x = 7 + b$
$x = 3$ $x = \frac{7 + b}{2}$ (done)

Problems

Solve each equation for the indicated variable.

1. $y + b = 7$ for y
2. $y + b = 7$ for b
3. $rt = 8$ for t

4. $ax = b$ for x
5. $2x + b = c$ for x
6. $2x + b = c$ for b

7. $3x - d = m$ for x
8. $17q = r$ for q
9. $a + t = b$ for t

10. $3x + 2y = 10$ for y
11. $5x + 2y = 8$ for x
12. $\pi d = c$ for d

Answers

1. $y = 7 - b$
2. $b = 7 - y$
3. $t = \frac{8}{r}$
4. $x = \frac{b}{a}$

5. $x = \frac{c - b}{2}$
6. $b = c - 2x$
7. $x = \frac{m + d}{3}$
8. $q = \frac{r}{17}$

9. $t = b - a$
10. $y = \frac{10 - 3x}{2}$
11. $x = \frac{8 - 2y}{5}$
12. $d = \frac{c}{\pi}$ for π

EQUATIONS WITH FRACTIONAL COEFFICIENTS

To remove a fraction multiplied by the variable, divide by the fraction or multiply by its reciprocal.

For additional information, see Year 2, Chapter 7, problem CT-41 on page 260. There are also additional examples of dividing fractions by fractions in the "Dividing Fractions Using Reciprocals" section of this guide on page 42.

Example 1

Solve the equation $\frac{2}{3}x = 8$.

Using division:

$$\frac{2}{3}x = 8$$

$$\frac{\frac{2}{3}x}{\frac{2}{3}} = \frac{8}{\frac{2}{3}}$$

$$x = 8 \div \frac{2}{3}$$

$$x = 8 \cdot \frac{3}{2}$$

$$x = 12$$

Using reciprocals:

$$\frac{2}{3}x = 8$$

$$\left(\frac{3}{2}\right)\frac{2}{3}x = 8\left(\frac{3}{2}\right)$$

$$1x = 12$$

$$x = 12$$

Example 2

Solve the equation $-\frac{1}{5}x - 1 = -3$.

Add 1 to isolate the variable.

Use division or reciprocals to solve.

$$-\frac{1}{5}x - 1 = -3$$

$$-\frac{1}{5}x = -2$$

$$x = -2 \div -\frac{1}{5} \text{ or } -2 \cdot (-5)$$

$$x = 10$$

Problems

Solve each equation.

1. $\frac{1}{5}x = 25$

2. $\frac{1}{3}x = 20$

3. $\frac{1}{4}x = 9$

4. $\frac{3}{4}x = 9$

5. $\frac{2}{3}x = \frac{1}{2}$

6. $-\frac{1}{2}x = 10$

7. $-\frac{5}{2}x = -10$

8. $1\frac{1}{3}x = 8$

9. $2\frac{1}{5}x = -33$

10. $\frac{1}{3}x - 7 = -3$

11. $\frac{1}{4}x + 2 = 6$

12. $-\frac{2}{3}x - 7 = -10$

Answers

1.   x = 125

2.   x = 60

3.   x = 36

4.   x = 12

5. $x = \frac{3}{4}$

6.   x = -20

7.   x = 4

8.   x = 6

9.   x = -15

10.   x = 12

11.   x = 16

12. $x = 4\frac{1}{2}$

GRAPHING

Graphing is a major thread of mathematics, so this section of the parent guide has many subtopics. The fundamentals of an xy-coordinate grid are taught in Year 1, Chapter 2, problems GS-24 through GS-27 on pages 42-43 or Year 2, Chapter 1, problem GO-66 on page 28.

SCALING THE AXES OF GRAPHS

The characteristics of a complete graph are listed in the Took Kit, Year 1, Chapter 1, problem AR-45 on page 17 and Year 2, Chapter 1, problem GO-10 on page 6. The axes of the graph must be marked with equal sized spaces called intervals. The numbers on the axes are the scaling of the axes. The difference between consecutive markings tells the size of each interval.

Sometimes the axis or set of axes is not provided. A student must count the number of usable spaces on the graph paper. How many spaces are usable depends in part on how large the graph will be and how much margin will be given for labeling beside each axis.

Follow these steps to scale each axis of a graph.

1. Find the difference between the smallest and largest numbers (the range) you need to put on an axis.
2. Count the number of intervals (spaces) you have on your axis.
3. Divide the range by the number of intervals to find the interval size.
4. Label the marks on the axis using the interval size.

Sometimes dividing the range by the number of intervals produces an interval size that makes it difficult to interpret the location of points on the graph. The student may exercise judgment and may round the interval size up (always up, if rounded at all) to a number that is convenient to use. Interval sizes like 1, 2, 5, 10, 20, 25, 50, 100, etc. work well.

Example 1

1. The difference between 0 to 60 is 60.
2. The number line is divided into 5 equal intervals.
3. 60 divided by 5 is 12
4. The marks are labeled with multiples of the interval size 12.

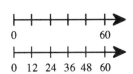

Example 2

1. The difference is 12 - (-16) = 28.
2. The number line is divided into 7 equal intervals.
3. 28 divided by 7 is 4
4. The marks are labeled with multiples of the interval size 4.

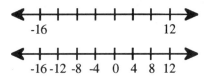

Example 3

1. The difference between 300 and 0 is 300.
2. There are 4 intervals
3. 300 ÷ 4 = 75
4. The axis is labeled with multiples of 75.

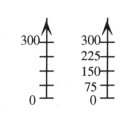

Example 4

1. The difference on the vertical axis is 750 – 0 = 750. (The origin is (0,0).) On the horizontal axis the range is 6 – 0 = 6.
2. There are 5 spaces vertically and 3 spaces horizontally.
3. The vertical interval size is 750 ÷ 5 = 150. The horizontal interval is 6 ÷ 3 = 2.
4. The axes are labeled appropriately.

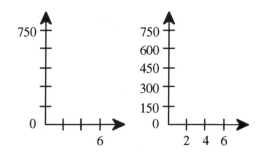

Example 5

Sometimes the axes extend in the negative direction. Use the positive direction origin to determine the interval size and then extend the intervals in the negative direction.

1. The vertical range 30 – 0 = 30. The horizontal range is 35 – 0 = 35
2. There are 5 intervals in each positive direction.
3. Vertical interval size is 30 ÷ 5 = 6. Horizontal interval size is 35 ÷ 5 = 7.
4. Label the axes.

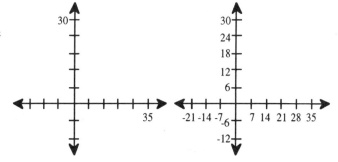

Problems

Scale each axis:

1.

2.

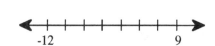

3.

150

70

4.

48

-12

5.

-13   -7

6.

-18   -6

7.

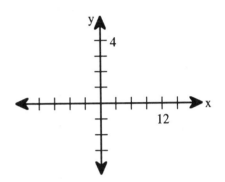
y
4

x
12

8.

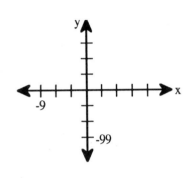
y

-9   x

-99

9.

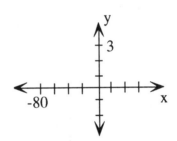
y
3

-80   x

10.

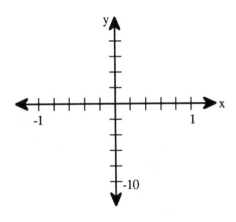
y

-1   1   x

-10

11.

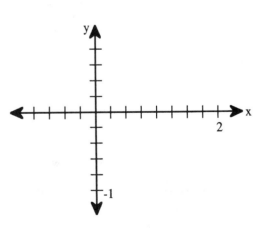
y

x
2

-1

12. Use fractions.

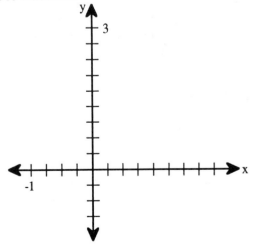
y
3

-1   x

Answers

1. 2, 4, 6, 8, 10, 12

2. -9, -6, -3, 0, 3, 6

3. 86, 102, 118, 134

4. -2, 8, 18, 28, 38

5. -12, -11, -10, -9, -8

6. -20, -16, -14, -12, -10, -8

7. x: -12, -9, -6, -3, 0, 3, 6, 9, 15
    y: -3, -2, -1, 0, 1, 2, 3

8. x: -9, -6, -3, 0, 3, 6, 9, 12
    y: -66, -33, 0, 33, 66, 99

9. x: -60, -40, -20, 0, 20, 40, 60
    y: -2, -1, 0, 1, 2

10. x: -0.8, -0.6, -0.4, -0.2, 0, 0.2, 0.4, 0.6, 0.8
     y: -8, -6, -4, -2, 0, 2, 4, 6, 8

11. x: -1, -0.75, -0.50, -0.25, -0, 0.25, 0.50, 0.75, 1.00, 1.25, 1.50, 1.75
     y: -0.8, -0.6, -0.4, -0.2, 0, 0.2, 0.4, 0.6, 0.8

12. x: $-\frac{3}{4}$, $-\frac{1}{2}$, $-\frac{1}{4}$, 0, $\frac{1}{4}$, $\frac{1}{2}$, $\frac{3}{4}$, 1, $\frac{5}{4}$, $\frac{3}{2}$, $\frac{7}{4}$, 2

     y: -1, $-\frac{2}{3}$, $-\frac{1}{3}$, 0, $\frac{1}{3}$, $\frac{2}{3}$, 1, $\frac{4}{3}$, $\frac{5}{3}$, 2, $\frac{7}{3}$, $\frac{8}{3}$

GRAPHING POINTS

Points on a coordinate grid are written as ordered pairs, (x, y), where the first number is the x-coordinate, that is, the horizontal distance from the y-axis. The second number is the y-coordinate, that is, the vertical distance from the x-axis. Taken together, the two coordinates name exactly one point on the graph. The examples below show how to place a point on an xy-coordinate grid. For more information, see Year 1, Chapter 2, problems GS-24 through 27 on pages 42-43 or Year 2, Chapter 1, problem GO-66 on page 28.

Example 1

Graph point A(2, -3).

Go right 2 units from the origin (0, 0), then go down 3 units. Mark the point.

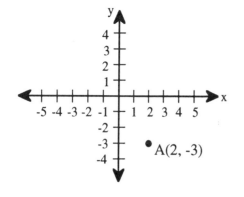

Example 2

Graph the ordered pair B(0, 3).

From the origin, do not go right or left; just go up 3 units. Mark the point.

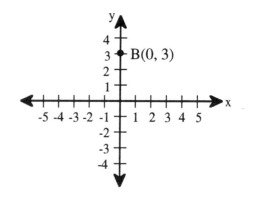

Example 3

Plot the point C(-4, 0) on a coordinate grid.

Go to the left from the origin 4 units, but do not go up or down. Mark the point.

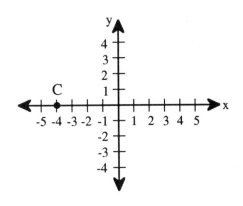

Example 4

Name the coordinate pair for each point on the graph.

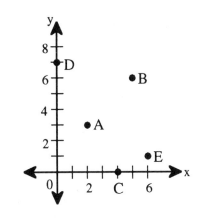

A(2, 3)   B(5, 6)   C(4, 0)   D(0, 7)   E(6, 1)

Problems

1. Name the coordinate pair for each point shown on the grid below.

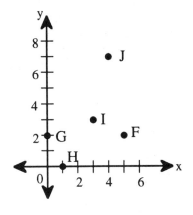

2. Use the ordered pair to locate each point on a coordinate grid.

K(0, -4)
L(-5, 0)
M(-2, -3)
N(-2, 3)
O(2, -3)
P(-4, -6)
Q(4, -5)
R(-5, -4)
T(-1, -6)

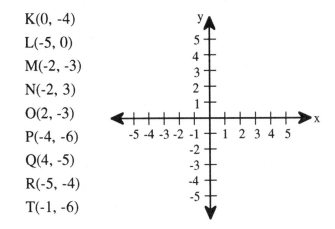

3. Name the coordinates of the points on the coordinate grid below.

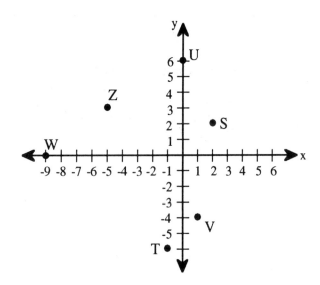

Answers

1. F(5, 2)
    G(0, 2)
    H(1, 0)
    I(3, 3)
    J(4, 7)

2.

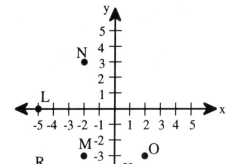

3. S(2, 2)
    T(-1, -6)
    U(0, 6)
    V(1, -4)
    W(-9, 0)
    Z(-5, 3)

GRAPHING ALGEBRAIC RELATIONSHIPS

An algebraic relationship is a rule that describes how two numbers are related. In the rule the numbers are usually represented by the variables x and y. More than one ordered pair (x, y) is described by each rule. To organize several ordered pairs for each rule, list them in a row or column or in a vertical or horizontal table. These organized tables are sometimes referred to as xy-tables or input/output tables. The ordered pairs are then used to make a graph that illustrates the relationship. For additional information, see Year 1, Chapter 2, problems GS-35 on page 45, GS-65 on page 53, and GS-67 on page 54 or Year 2, Chapter 2, problems FT-19 on page 46 and FT-45 on page 54.

Example 1

Graph the algebraic relationship $y = x + (-2)$.

Students need to find and organize some ordered pairs for the graph. Start with a table of x-values. Then take the x-values (or the input values) in the table and add -2 to each of them to get the corresponding y-value (output).

$$-2 + (-2) = -4$$
$$-1 + (-2) = -3$$
$$1 = (-2) = -1$$

Once the (x, y) values are completed in the table, place those points on the coordinate grid. Remember to start at the origin (0,0) and move horizontally right or left before going up or down for the second number (coordinate).

Those points are written (x, y): (-3, -5), (-2, -4), (-1, -3)

Finally, draw the line or curve through the points when the input values may be any number, including fractions, decimals, and irrational numbers.

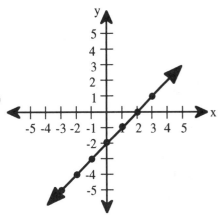

Example 2

For $y = x + 2$, add 2 to each x-value to get the corresponding y-value.

x	-3	-2	-1	0	1	2	3
y	-1	0	1	2	3	4	5

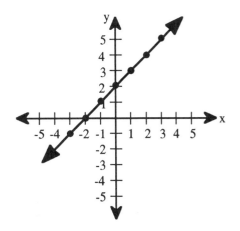

Example 3

For $y = 2x$, multiply each x-value to get its corresponding y-value.

x	-3	-2	-1	0	1	2	3
y	-6	-4	-2	0	2	4	6

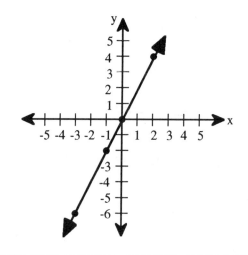

You may want to use the same x values for each problem. However, some rules are easier to graph if you select x-values that make the arithmetic simpler. For example, if you use even numbers in problem 5 below, all y-values will be integers.

x	-3	-2	-1	0	1	2	3
y							

Problems

For each rule, make a table to organize the ordered pairs, then graph the relationship on a coordinate grid.

1. $y = x + (-1)$ 2. $y = x + (-4)$ 3. $y = 2x + 2$ 4. $y = 3x + 1$

5. $y = \frac{x}{2}$ 6. $y = \frac{x}{2} + 3$ 7. $y = \frac{x}{3}$ 8. $y = 3x - 3$

Answers

1. $y = x + (-1)$

x	-3	-2	-1	0	1	2	3
y	-4	-3	-2	-1	0	1	2

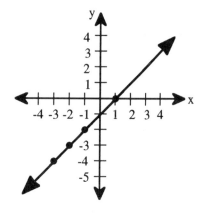

2. $y = x + (-4)$

x	-3	-2	-1	0	1	2	3
y	-7	-6	-5	-4	-3	-2	-1

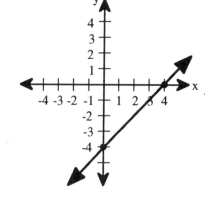

3. $y = 2x + 2$

x	-3	-2	-1	0	1	2	3
y	-4	-2	0	2	4	6	8

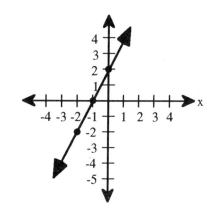

Algebra and Functions

4. $y = 3x + 1$

x	-4	-2	-1	0	4
y	-11	-5	-2	1	13

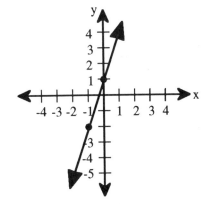

5. $y = \frac{x}{2}$

x	-4	-2	-1	0	4
y	-2	-1	-.5	0	2

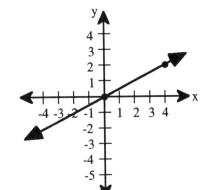

6. $y = \frac{x}{2} + 3$

x	-4	-2	-1	0	4
y	1	2	2.5	3	5

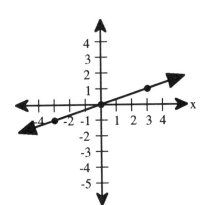

7. $y = \frac{x}{3}$

x	-3	-1	0	1	3
y	-1	$-\frac{1}{3}$	0	$\frac{1}{3}$	1

8. $y = 3x - 3$

x	-4	-2	-1	0	4
y	-15	-9	-6	-3	9

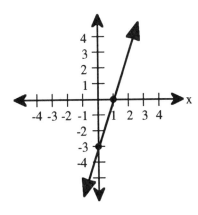

The relationship described by each of these rules is a line. Other algebraic rules may produce graphs which follow patterns that are not lines but curves.

WRITING AND GRAPHING LINEAR EQUATIONS ON A FLAT SURFACE

Slope is a number that indicates the steepness (or flatness) of a line, as well as its direction (up or down) left to right.

Slope is determined by the ratio $\dfrac{\text{vertical change}}{\text{horizontal change}}$ between <u>any</u> two points on a line.

For lines that go up (from left to right), the sign of the slope is positive. For lines that go down (left to right), the sign of the slope is negative.

Any linear equation written as $y = mx + b$, where m and b are any real numbers, is said to be in slope-intercept form. The slope of the line is m. The y-intercept is b, that is, the point (0, b) where the line intersects (crosses) the y-axis.

For additional information, see Year 2, Chapter 9, problem CB-39 on page 344.

Example 1

Write the slope of the line containing the points (-1, 3) and (4, 5).

First graph the two points and draw the line through them.

Look for and draw a slope triangle using the two given points.

Write the ratio $\dfrac{\text{vertical change in } y}{\text{horizontal change in } x}$ using the legs of the right triangle: $\dfrac{2}{5}$.

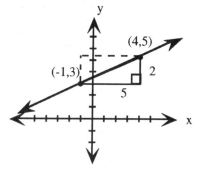

Assign a positive or negative value to the slope (this one is positive) depending on whether the line goes up (+) or down (−) from left to right.

If the points are inconvenient to graph, use a "Generic Slope Triangle," visualizing where the points lie with respect to each other.

Example 2

Graph the linear equation $y = \dfrac{4}{7}x + 2$ without making a table.

Using $y = mx + b$, the slope in $y = \dfrac{4}{7}x + 2$ is $\dfrac{4}{7}$ and the y-intercept is the point (0, 2). To graph, begin at the y-intercept (0, 2). Remember that slope is $\dfrac{\text{vertical change}}{\text{horizontal change}}$ so go up 4 units (since 4 is positive) from (0, 2) and then move right 7 units. This gives a second point on the graph. To create the graph, draw a straight line through the two points.

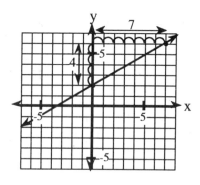

Problems

Write the slope of the line containing each pair of points.

1. (3, 4) and (5, 7)

2. (5, 2) and (9, 4)

3. (1, -3) and (-4, 7)

4. (-2, 1) and (2, -2)

5. (-2, 3) and (4, 3)

6. (8, 3) and (3, 5)

Identify the slope and y-intercept in each equation.

7. $y = \frac{1}{2}x - 2$

8. $y = -\frac{3}{5}x - \frac{5}{3}$

9. $y = 2x + 4$

10. $y = \frac{2}{3}x - 5$

11. $y = -3x - \frac{1}{2}$

12. $\frac{1}{2}x + 3$

Determine the slope of each line using the <u>highlighted</u> <u>points</u>.

13.

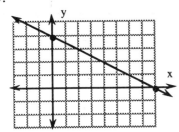

14.

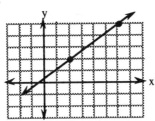

15.

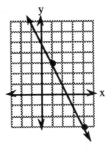

Graph the following linear equations on graph paper without making a table.

16. $y = \frac{1}{2}x + 3$

17. $y = -\frac{3}{5}x - 1$

18. $y = 4x$

19. $y = -6x + \frac{1}{2}$

20. $3x + 2y = 12$

Answers

1. $\frac{3}{2}$

2. $\frac{1}{2}$

3. - 2

4. $-\frac{3}{4}$

5. 0

6. $-\frac{2}{5}$

7. $\frac{1}{2}$, (0, -2)

8. $-\frac{3}{5}$, $(0, -\frac{5}{3})$

9. 2, (0, 4)

10. $\frac{2}{3}$, (0, -5)

11. -3, $(0, -\frac{1}{2})$

12. $\frac{1}{2}$, (0, 3)

13. $-\frac{1}{2}$

14. $\frac{3}{4}$

15. -2

16. line with slope $\frac{1}{2}$ and y-intercept (0, 3)

17. line with slope $-\frac{3}{5}$ and y-intercept $(0, -1)$ 18. line with slope 4 and y-intercept $(0, 0)$

19. line with slope -6 and y-intercept $(0, \frac{1}{2})$ 20. line with slope $-\frac{3}{2}$ and y-intercept $(0, 6)$

Note: The topic of rate of change is addressed in the Proportions section of this guide (found in "Number Sense," p. 21).

INEQUALITIES

GRAPHING INEQUALITIES

The solutions to an equation can be represented as a point (or points) on the number line. The solutions to inequalities are represented by rays or segments with solid or open endpoints. Solid endpoints indicate that the endpoint is included in the solution ($\leq$ or $\geq$), while the open dot indicates that it is not part of the solution ($<$ or $>$). For additional information, see year 2, Chapter 9, problems CB-79 and 80 on pages 354-55 and CB-92 on page 358.

Example 1

$x > 6$

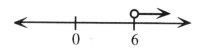

Example 2

$x \leq -1$

Example 3

$-1 \leq y < 6$

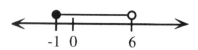

Example 4

$y \geq -2$

Problems

Graph each inequality on a number line.

1. $m < 2$

2. $x \leq -1$

3. $y \geq 3$

4. $-1 \leq x \leq 3$

5. $-6 < x < -2$

6. $-1 < x \leq 2$

7. $m > -9$

8. $x \neq 1$

9. $x \leq 3$

Answers

1.
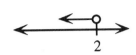

2.

3.
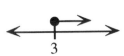

4.
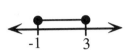

5.
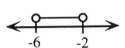

6.
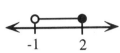

7.
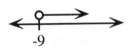

8.
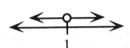

9.
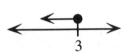

SOLVING INEQUALITIES

To solve an inequality, first solve it like any equation. Use the solution as a dividing point of the line. Then test a value from each side of the dividing point on the number line. If the test number is true, then that part of the number line is part of the solution. In addition, if the inequality is ≥ or ≤, then the dividing point is part of the solution and is indicated by a solid dot. If the inequality is > or <, then the dividing point is not part of the solution, indicated by an open dot.

Example 1

$9 \geq m + 2$

Solve the equation:
$$9 = m + 2$$
$$7 = m$$

Draw a number line. Put a solid dot at 7.

Test a number on each side of 7 in the original inequality. We use 10 and 0.

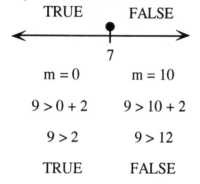

TRUE	FALSE
m = 0	m = 10
$9 > 0 + 2$	$9 > 10 + 2$
$9 > 2$	$9 > 12$
TRUE	FALSE

The solution is $m \leq 7$.

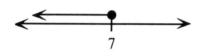

Example 2

$-2x - 3 < x + 6$

Solve the equation:
$$-2x - 3 = x + 6$$
$$-2x = x + 9$$
$$-3x = 9$$
$$x = -3$$

Draw a number line. Put an open dot at -3.

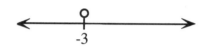

Test 0 and -4 in the original inequality.

FALSE	TRUE
x = -4	x = 0
$-2(-4) - 3 < -4 + 6$	$-2(0) - 3 < 0 + 6$
$8 - 3 < 2$	$-3 < 6$
$5 < 2$	TRUE
FALSE	

The solution is $x > -3$.

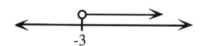

Problems

Solve each inequality.

1. $x + 3 > -1$
2. $y - 3 \leq 5$
3. $-3x \leq -6$
4. $2m + 1 \geq -7$
5. $-7 < -2y + 3$
6. $8 \geq -2m + 2$
7. $2x - 1 < -x + 8$
8. $2(m + 1) \geq m - 3$
9. $3m + 1 \leq m + 7$

Answers

1. $x > -4$
2. $y \leq 8$
3. $x \geq 2$
4. $m \geq -4$
5. $y < 5$
6. $m \geq -3$
7. $x < 3$
8. $m \geq -5$
9. $m \leq 3$

Measurement

and

Geometry

Area

Perimeter of Polygons and Circumference of Circles

Surface Area

Volume

The Pythagorean Theorem

Angles, Triangles, and Quadrilaterals

Scale Factor and Ratios of Growth

Area is the number of square units in the interior region of a planc (flat) figure (two dimensional) or the surface area of a three-dimensional figure. For example, area is the region that is covered by floor tile (two dimensional) or paint on a box or ball (three dimensional).

AREA OF A RECTANGLE

For the Tool Kit entry about area of a rectangle, see Year 1, Chapter 4, problem MP-16 on page 105. To find the area of a rectangle, follow the steps below.

1. Identify the base.
2. Identify the height.
3. Multiply the base times the height to find the area in square units: **A = bh**.

A square is a rectangle in which the base and height are of equal length. Find the area of a square by multiplying the base times itself: **A = b²**.

Example

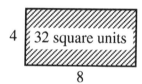

base = 8 units

height = 4 units

A = 8 · 4 = 32 square units

Problems

Find the areas of the rectangles (problems 1-8) and squares (problems 9-12) below.

1.
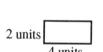
2 units
4 units

2.
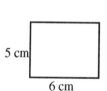
5 cm
6 cm

3.

7 in.
3 in.

4.
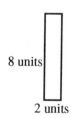
8 units
2 units

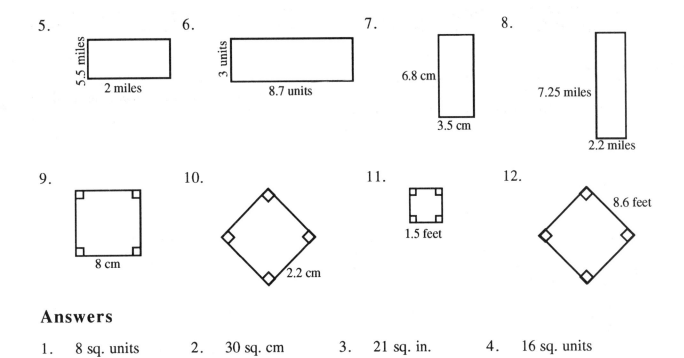

5. 5.5 miles, 2 miles

6. 3 units, 8.7 units

7. 6.8 cm, 3.5 cm

8. 7.25 miles, 2.2 miles

9. 8 cm

10. 2.2 cm

11. 1.5 feet

12. 8.6 feet

Answers

1. 8 sq. units	2. 30 sq. cm	3. 21 sq. in.	4. 16 sq. units
5. 11 sq. miles	6. 26.1 sq. units	7. 23.8 sq. cm	8. 15.95 sq. miles
9. 64 sq. cm	10. 4.84 sq. cm	11. 2.25 sq. feet	12. 73.96 sq. feet

AREA OF A PARALLELOGRAM

A parallelogram is easily changed to a rectangle by separating a triangle from one end of the parallelogram and moving it to the other end as shown in the three figures below. For additional information, see Year 1, Chapter 4, problems MP-26 on page 108 and MP-43 on page 114 or Year 2, Chapter 6, problems RS-33 on page 216 and RS-52 on page 221.

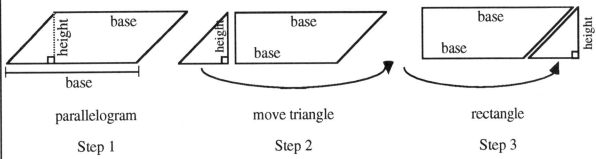

parallelogram
Step 1

move triangle
Step 2

rectangle
Step 3

To find the area of a parallelogram, multiply the base times the height: **A = bh**. Notice that the height is <u>not</u> the same as the second side of the parallelogram (as it was in the rectangle).

Example

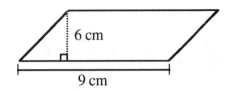

base = 9 cm

height = 6 cm

A = 9 · 6 = 54 square cm

Problems

Find the area of each parallelogram below.

1.

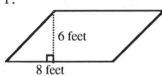

6 feet

8 feet

2.

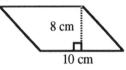

8 cm

10 cm

3.

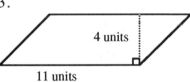

4 units

11 units

4.

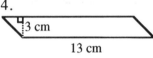

3 cm

13 cm

5.

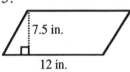

7.5 in.

12 in.

6.

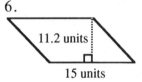

11.2 units

15 units

7.

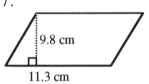

9.8 cm

11.3 cm

8.

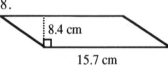

8.4 cm

15.7 cm

Answers

1.   48 sq. feet   2.   80 sq. cm   3.   44 sq. units   4.   39 sq. cm

5.   90 sq. in.   6.   168 sq. units 7.   110.74 sq. cm   8.   131.88 sq. cm

AREA OF A TRIANGLE

The area of a triangle is equal to one half the area of a parallelogram. This fact can easily be shown by cutting a parallelogram in half along a diagonal (see below). For additional information, see Year 1, Chapter 4, problem MP-55 on page 117 or Year 2, Chapter 6, problem RS-52 on page 221.

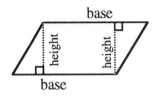

parallelogram

Step 1

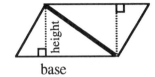

draw a diagonal

Step 2

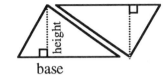

match triangles by cutting apart or by folding

Step 3

By cutting or folding the parallelogram along the diagonal, the result is two congruent (same size and shape) triangles. Thus, the area of a triangle has half the area of the parallelogram that can be created from two copies of the triangle.

To find the area of a triangle, follow the steps below.

1. Identify the base.
2. Identify the height.
3. Multiply the base times the height.
4. Divide the product of the base times the height by 2: $A = \dfrac{bh}{2}$ or $\dfrac{1}{2} bh$.

Example 1

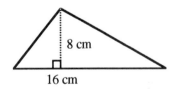

base = 16 cm

height = 8 cm

$A = \dfrac{16 \cdot 8}{2} = \dfrac{128}{2} = 64 \text{ cm}^2$

Example 2

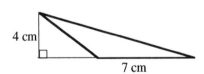

base = 7 cm

height = 4 cm

$A = \dfrac{7 \cdot 4}{2} = \dfrac{28}{2} = 14 \text{ cm}^2$

Problems

1.
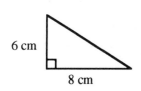

2.
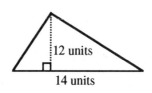

3.
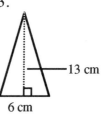

4.
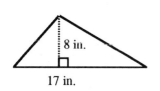

5.
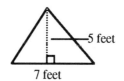

6.
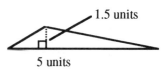

7.
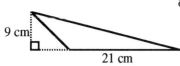

8.
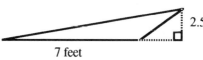

Answers

1. 24 sq. cm
2. 84 sq. units
3. 39 sq. cm
4. 68 sq. in.

5. 17.5 sq. feet
6. 3.75 sq. units
7. 94.5 sq. cm
8. 8.75 sq. feet

Measurement and Geometry

AREA OF A TRAPEZOID

A trapezoid is another shape that can be transformed into a parallelogram. Change a trapezoid into a parallelogram by following the three steps below.

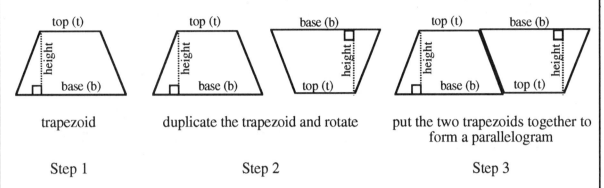

trapezoid	duplicate the trapezoid and rotate	put the two trapezoids together to form a parallelogram
Step 1	Step 2	Step 3

To find the area of a trapezoid, multiply the base of the large parallelogram in step 3 (base and top) times the height and then take half of the total area. Remember to add the lengths of the base and the top of the trapezoid before multiplying by the height. Note that some texts call the top length the upper base and the base the lower base.

$$A = \frac{1}{2}(b + t)\ h \quad \text{or} \quad \frac{b + t}{2} \cdot h$$

For additional information, see Year 1, Chapter 4, problem MP-109 on page 134 or Year 2, Chapter 6, problem RS-52 on page 221.

Example

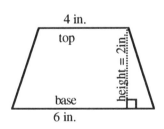

top = 4 in.

base = 6 in.

height = 2 in.

$$A = \frac{4 + 6}{2} \cdot 2 = \frac{10}{2} \cdot 2 = 5 \cdot 2 = 10 \text{ in.}^2$$

Problems

Find the areas of the trapezoids below.

1.

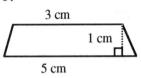

3 cm

1 cm

5 cm

2.

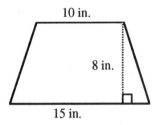

10 in.

8 in.

15 in.

3.

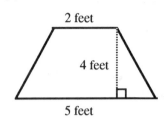

2 feet

4 feet

5 feet

4.

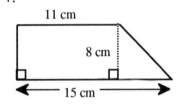

11 cm

8 cm

15 cm

5.

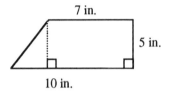

7 in.

5 in.

10 in.

6.

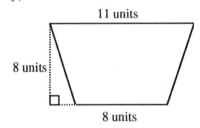

11 units

8 units

8 units

7.

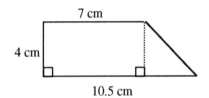

7 cm

4 cm

10.5 cm

8.

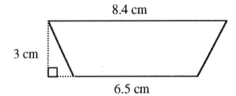

8.4 cm

3 cm

6.5 cm

Answers

1.   4 sq. cm

2.   100 sq. in.

3.   14 sq. feet

4.   104 sq. cm

5.   42.5 sq. in.

6.   76 sq. units

7.   35 sq. cm

8.   22.35 sq. cm

CALCULATING COMPLEX AREAS USING SUBPROBLEMS

Students can use their knowledge of areas of polygons to find the areas of more complicated figures. The use of subproblems (that is, solving smaller problems in order to solve a larger problem) is one way to find the areas of complicated figures. For additional information, see Year 2, Chapter 8, problems GS-43 and 45 on pages 303-04.

Example 1

Find the area of the figure at right.

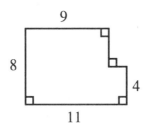

Method #1	Method #2	Method #3

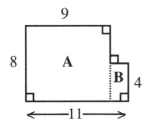   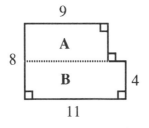   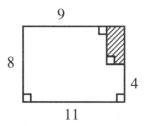

Subproblems:	Subproblems:	Subproblems:

1. Find the area of rectangle A:

   $8 \cdot 9 = 72$ square units

2. Find the area of rectangle B:

   $4 \cdot (11 - 9) = 4 \cdot 2 = 8$ square units

3. Add the area of rectangle A to the area of rectangle B:

   $72 + 8 = 80$ square units

1. Find the area of rectangle A:

   $9 \cdot (8 - 4) = 9 \cdot 4 = 36$ square units

2. Find the area of rectangle B:

   $11 \cdot 4 = 44$ square units

3. Add the area of rectangle A to the area of rectangle B:

   $36 + 44 = 80$ square units

1. Make a large rectangle by enclosing the upper right corner.

2. Find the area of the new, larger rectangle:

   $8 \cdot 11 = 88$ square units

3. Find the area of the shaded rectangle:

   $(8 - 4) \cdot (11 - 9)$
   $= 4 \cdot 2 = 8$ square units

4. Subtract the shaded rectangle from the larger rectangle:

   $88 - 8 = 80$ square units

Example 2

Find the area of the figure at right.

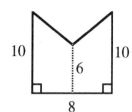

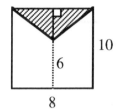

Subproblems:

1. Make a rectangle out of the figure by enclosing the top.

2. Find the area of the entire rectangle: $8 \cdot 10 = 80$ square units

3. Find the area of the shaded triangle. Use the formula $A = \frac{1}{2}$ bh.

   $b = 8$ and $h = 10 - 6 = 4$, so $A = \frac{1}{2} (8 \cdot 4) = \frac{32}{2} = 16$ square units

4. Subtract the area of the triangle from the area of the rectangle:
   $80 - 16 = 64$ square units

Problems

Find the areas of the complex figures below.

1.

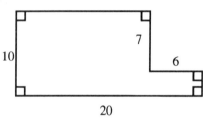

2.

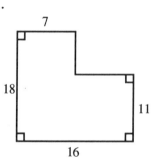

3.

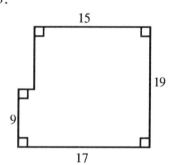

4.

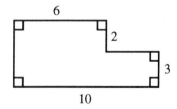

5.

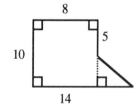

6. The slanted segments have equal lengths.

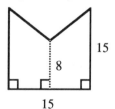

7.

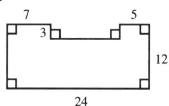

8.

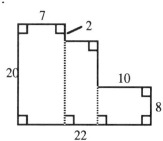

9.

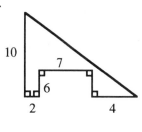

10. The slanted segments have equal length.

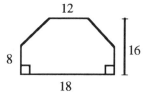

11. Find the area of the shaded region bounded by a rectangle and triangle.

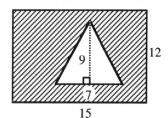

12. Find the area of the shaded region bounded by the two rectangles.

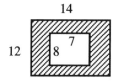

Answers

1.	158 sq. units	2.	225 sq. units	3.	303 sq. units	4.	42 sq. units
5.	95 sq. units	6.	172.5 sq. units	7.	252 sq. units	8.	310 sq. units
9.	23 sq. units	10.	264 sq. units	11.	148.5 sq. units	12.	112 sq. units

AREA OF A CIRCLE

In class, students have done explorations with circles and circular objects to discover the relationship between circumference, diameter, and pi (π). To read more about the in-class exploration see Year 1, Chapter 9, problems ZC-10, 20, 21, 23, and 24 as well as the Tool Kit entry, Chapter 9, problem ZC-25 on pages 326-31. In Year 2, Chapter 6, see problem RS-61 as well as the Tool Kit entry, problem RS-63 on pages 224-25.

In order to find the area of a circle, identify the radius of the circle. The radius is half the diameter. Next, square the radius and multiply the result by π. Depending on the teacher's or book's preference, students may use $\frac{22}{7}$ for π when the radius or diameter is a fraction, 3.14 for π as an approximation, or the π button on the calculator. When using the π button, most teachers will want students to round to the nearest tenth or hundredth.

The formula for the area of a circle is: $\mathbf{A = \pi r^2}$.

Note: all of the examples and answers in this book will use $\pi \approx 3.14$.

Example 1

Find the area of a circle with $r = 17$ feet.

$A = \pi(17)^2$
$\approx 3.14(17 \cdot 17)$
≈ 907.46 square feet

Example 2

Find the area of a circle with $d = 84$ cm.

$r = 42$ cm
$A = \pi(42)^2$
$\approx 3.14(42 \cdot 42)$
≈ 5538.96 square cm

Example 3

Find the radius of a circle with area 78.5 square meters.

$78.5 = \pi r^2$
$78.5 = 3.14 r^2$
$\dfrac{78.5}{3.14} = \dfrac{3.14}{3.14} r^2$
$78.5 \div 3.14 = 25$

$25 = r^2$

$r = \sqrt{25}$
$= 5$ meters

Example 4

Find the radius of a circle with area 50.24 square centimeters.

$50.24 = \pi r^2$
$50.24 = 3.14 r^2$
$\dfrac{50.24}{3.14} = \dfrac{3.14}{3.14} r^2$
$16 = r^2$

$r = \sqrt{16}$
$= 4$ centimeters

Problems

Find the area of circles with the following radius or diameter lengths. Remember, use 3.14 for π. Round to the nearest hundredth.

1. $r = 6$ cm

2. $r = 3.2$ in.

3. $d = 16$ ft

4. $r = \frac{1}{2}$ m

5. $d = \frac{4}{5}$ cm

6. $r = 5$ in.

7. $r = 3.6$ cm

8. $r = 2\frac{1}{4}$ in.

9. $d = 14.5$ ft

10. $r = 12.02$ m

Find the radius of each circle given the following areas. Round answers to the nearest tenth.

11. $A \approx 36.29$ m^2

12. $A \approx 63.59$ cm^2

13. $A = 153.86$ ft^2

14. $A = 530.66$ in^2

15. $A = 415.265$ km^2

Answers

1. 113.04 cm^2

2. 32.15 in^2

3. 200.96 ft^2

4. 0.79 m^2

5. 0.50 cm^2

6. 78.5 in^2

7. 40.69 cm^2

8. 15.90 in^2

9. 165.05 ft^2

10. 453.67 m^2

11. $r \approx 3.4$ m

12. $r \approx 4.5$ cm

13. $r = 7$ ft

14. $r = 13$ in.

15. $r = 11.5$ km

SUMMARY TABLE FOR AREA

Shape	Formula	Example
rectangle	$A = bh$	3 cm ▭ 6 cm $= 6 \cdot 3 = 18$ square cm
square	$A = b^2$	2 cm □ 2 cm $= 2 \cdot 2 = 4$ square cm
parallelogram	$A = bh$	6 cm, 10 cm $= 10 \cdot 6 = 60$ square cm
triangle	$A = \dfrac{bh}{2}$ or $\dfrac{1}{2}bh$	4 cm, 7 cm $= \dfrac{7 \cdot 4}{2} = \dfrac{28}{2} = 14$ square cm
trapezoid	$A = \dfrac{1}{2}(b + t)h$ or $\dfrac{b + t}{2} \cdot h$	3 cm, 5 cm, 7 cm $= \dfrac{7 + 3}{2} \cdot 5 = 25$ square cm
circle	$A = \pi r^2$	6 cm $= 3.14(6)^2 = 113.04$ square cm

PERIMETER

The perimeter of a polygon is the distance around the outside of the figure. The perimeter is found by adding the lengths of all of the sides. The perimeter is similar to a fence put around a yard.

In some of the examples below, a subproblem is needed to find the length of the missing side(s) before finding the perimeter.

Example 1

Find the perimeter of the parallelogram below.

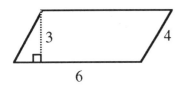

P = 6 + 4 + 6 + 4 = 20 units

(Parallelograms have opposite sides equal.)

Example 2

Find the perimeter of the triangle below.

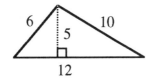

P = 6 + 10 + 12 = 28 units

Example 3

Find the perimeter of the figure below.

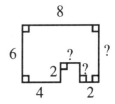

P = 6 + 8 + 6 + 2 + 2 + 2 + 2 + 4
    = 32 units

(You need to look carefully to find the lengths of the missing sides.)

Note: In some problems the height might be needed to do subproblems such as finding the length of the side of the figure. Since all of the sides are known in examples 1 and 2 above, the height is not used in the calculations.

Problems

Find the perimeter of each shape.

1. a rectangle with b = 5 and h = 10
2. a square with sides of length 9
3. a parallelogram with b = 8 and s = 5
4. a triangle with sides 4, 10, and 12

5. a parallelogram

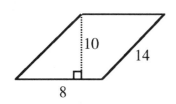

6.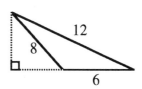

7.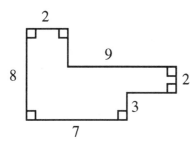

8.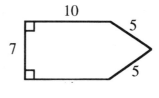

9.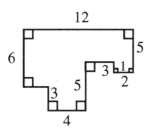

10.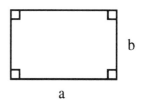

Answers

1. 30 units
2. 36 units
3. 26 units
4. 26 units

5. 44 units
6. 26 units
7. 38 units
8. 37 units

9. 44 units
10. a + b + a + b = 2a + 2b

CIRCUMFERENCE

The circumference of a circle is similar to the perimeter of a polygon. The circumference is the length of a circle. The circumference would tell you how much string it would take to go around a circle once.

Circumference is explored by students in Year 1, Chapter 9, problem ZC-10 and Tool Kit entry ZC-11 on pages 326-27 or in Year 2, Chapter 6, problem RS-57 and Tool Kit entry RS-62 on pages 222-25. The ratio of the circumference to the diameter of a circle is pi (π). Circumference is found by multiplying π by

the diameter. Students may use $\frac{22}{7}$, 3.14, or the π button on their calculator, depending on the teacher's or the book's directions.

$$C = 2\pi r \quad \text{or} \quad C = \pi d$$

For additional information, see Year 1, Chapter 9, problem ZC-11 on pages 326-27 or Year 2, Chapter 5, problem RS-62 on page 225.

Example 1

Find the circumference of a circle with a diameter of 5 inches.

d = 5 inches

$C = \pi d$
$= \pi(5)$ or 3.14(5)
≈ 15.7 inches

Example 2

Find the circumference of a circle with a radius of 10 units.

r = 10, so d = 2(10) = 20

$C = \pi 20$
$= 3.14(20)$
≈ 62.8 units

Example 3

Find the diameter of a circle with a circumference of 163.28 inches.

$C = \pi d$
$163.28 = \pi d$
$163.28 \approx 3.14d$
$d \approx \frac{163.28}{3.14}$

≈ 52 inches

Problems

Find the circumference of each circle given the following radius or diameter lengths. Remember to use 3.14 as the approximation for π. Round your answer to the nearest hundredth.

1. d = 12   2. d = 3.4   3. r = 2.1   4. d = 25   5. r = 1.54

Find the circumference of each circle shown below. Round your answer to the nearest hundredth.

6.

7.

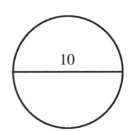

Find the diameter of each circle given the circumference. Round your answer to the nearest tenth.

8. C = 48.36   9. C = 35.6   10. C = 194.68

Answers

1. 37.68 units   2. 10.68 units   3. 13.19 units   4. 78.5 units

5. 9.67 units   6. 25.12 units   7. 31.4 units   8. 15.4 units

9. 11.34 units   10. 62 units

SURFACE AREA

SURFACE AREA OF A PRISM

The surface area of a prism (SA) is the sum of the areas of all of the faces, including the bases. Surface area is expressed in square units.

For additional information, see Year 1, Chapter 9, problems ZC-84 and 87 on pages 346-48 or Year 2, Chapter 8 problems GS-90 on pages 315-16.

Example

Find the surface area of the triangular prism at right.

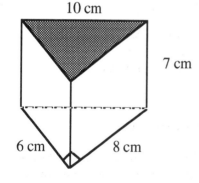

Subproblem 1: Area of the 2 bases

$$2\left[\tfrac{1}{2}(6\text{ cm})(8\text{ cm})\right] = 48\text{ cm}^2$$

Subproblem 2: Area of the 3 sides (lateral faces)

Area of left face: (6 cm)(7 cm) = 42 cm²
Area of right face: (8 cm)(7 cm) = 56 cm²
Area of back face: (10 cm)(7 cm) = 70 cm²

Subproblem 3: Surface Area of Prism = sum of bases and lateral faces

SA = 48 cm² + 42 cm² + 56 cm² + 70 cm² = 216 cm²

Problems

Find the surface area of each prism. All of the lateral faces are rectangles.

1.

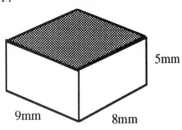

5mm

9mm   8mm

2.

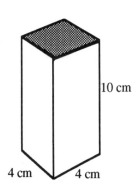

10 cm

4 cm   4 cm

3.

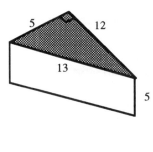

5   12

13

5

4.

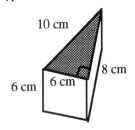

10 cm

8 cm

6 cm   6 cm

5. The pentagon is equilateral.

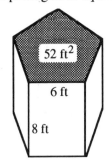

52 ft²

6 ft

8 ft

6.

2 cm

10 cm

6 cm   6 cm

10 cm

Answers

1.   314 mm²

2.   192 cm²

3.   210 units²

4.   192 cm²

5.   344 ft²

6.   408 cm²

SURFACE AREA OF A CYLINDER

The surface area of a cylinder is the sum of the two base areas and the lateral surface area. The formula for the surface area is:

$$SA = 2\pi r^2 + \pi dh \quad \text{or} \quad SA = 2\pi r^2 + 2\pi rh$$

where r = radius, d = diameter, and h = height of the cylinder. For additional information, see Year 1, Chapter 9, problem ZC-87 on page 348 or Year 2, Chapter 8, problem GS-123 on page 323.

Example 1

Find the surface area of the cylinder at right. Use 3.14 for π.

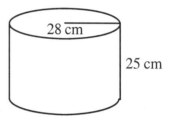

Subproblem 1: Area of the two circular bases

$$2[\pi(28 \text{ cm})^2] = 1568\pi \text{ cm}^2$$

Subproblem 2: Area of the lateral face (a rectangle with one side equal to the circumference of each base)

$$\pi(56)25 = 1400\pi \text{ cm}^2$$

Subproblem 3: Surface area of the cylinder

$$1568\pi \text{ cm}^2 + 1400\pi \text{ cm}^2 = 2968\pi \text{ cm}^2$$
$$\approx 2968(3.14) \text{ cm}^2$$
$$\approx 9319.32 \text{ cm}^2$$

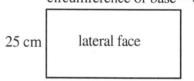

circumference of base = 56π cm

Example 2

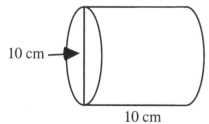

$$\begin{aligned}
SA &= 2\pi r^2 + 2\pi rh \\
&= 2\pi (5)^2 + 2\pi \cdot 5 \cdot 10 \\
&= 50\pi + 100\pi \\
&= 150\pi \\
&\approx 150(3.14) \\
&\approx 471 \text{ cm}^2
\end{aligned}$$

Example 3

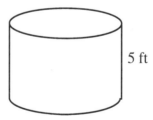

If the volume of the tank above is 1571.80 ft³, what is the surface area? Use 3.14 for π.

$$\begin{aligned}
V &= \pi r^2 h \\
1570 &= \pi r^2(5) \\
\frac{1570}{5\pi} &= r^2 \\
100 &= r^2 \\
10 &= r
\end{aligned}$$

$$\begin{aligned}
SA &= 2\pi r^2 + 2\pi rh \\
&= 2\pi 10^2 + 2\pi(10)(5) \\
&= 200\pi + 100\pi \\
&= 300\pi \approx 942 \text{ ft}^2
\end{aligned}$$

Problems

Find the surface area of each cylinder. Use 3.14 for π. Round your answers to the nearest hundredth

1. $r = 6$ cm, $h = 10$ cm
2. $r = 3.5$ in., $h = 25$ in.
3. $d = 9$ in., $h = 8.5$ in.

4. $d = 15$ cm, $h = 10$ cm
5. base area = 25 ft. height = 8 ft.
6. Volume = 1000 cm³, height = 25 cm

Answers

1. 602.88 cm²
2. 626.43 in²
3. 367.38 in²

4. 824.25 cm²
5. 191.76 ft.²
6. 640.36 cm²

VOLUME

VOLUME OF A PRISM

Volume is a three-dimensional concept. It measures the amount of interior space of a three-dimensional figure based on a cubic unit, that is, the number of 1 by 1 by 1 cubes that will fit inside a figure. In this textbook series students will calculate the volume of prisms, cylinders, and cones.

The volume of a prism is the area of either base (A) times the height (h) of the prism.

V = (Area of base) · (height) or V = Ah

For additional information, see Year 1, Chapter 9, problem ZC-58 on page 339 or Year 2, Chapter 8, problem GS-79 on page 312.

Example 1

Find the volume of the square prism below.

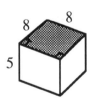

The base is a square with area (A) $8 \cdot 8 = 64$ units².

Volume = A(h)
= 64(5)
= 320 units³

Example 2

Find the volume of the triangular prism below.

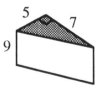

The base is a right triangle with area $\frac{1}{2}(5)(7) \approx 17.5$ units².

Volume = A(h)
= 17.5(9)
= 157.5 units³

Example 3

Find the volume of the trapezoidal prism below.

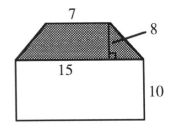

The base is a trapezoid with

area $\frac{1}{2}(7 + 15) \cdot 8 = 88$ units^2.

Volume $= A(h)$
$= 88(10)$
$= 880$ units^3

Example 4

Find the height of the prism with a volume of 132.5 cm^3 and base area of 25 cm^2.

Volume $= A(h)$
$132.5 = 25(h)$
$h = \frac{132.5}{25}$
$h = 5.3$ cm

Problems

Calculate the volume of each prism. The base of each figure is shaded. All lateral faces are rectangles.

1. Rectangular Prism

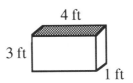

2. Right Triangular Prism

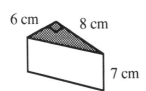

3. Rectangular Prism.

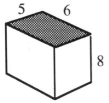

4. Right Triangular Prism

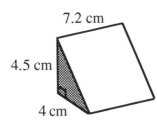

5. Trapezoidal Prism

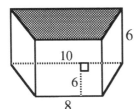

6. Triangular Prism with
    $A = 15\frac{1}{2}$ units^2

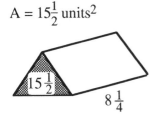

7. Find the volume of a prism with base area 32 cm^2 and height 1.5 cm.

8. Find the height of a prism with base area 32 cm^2 and volume 176 cm^3.

9. Find the base area of a prism with volume 47.01 cm^3 and height 3.2 cm.

Answers

1. 12 ft³   2. 168 cm³   3. 240 units³   4. 64.8 cm³   5. 324 units³

6. $127\frac{7}{8}$ units³ 7. 48 cm³   8. 5.5 cm   9. 14.69 cm²

VOLUME OF A CYLINDER

The volume of a cylinder is the area of its base multiplied by its height:

$$V = A \cdot h.$$

Since the base of a cylinder is a circle of area $A = \pi r^2$, we can write:

$$V = \pi r^2 h.$$

For additional information, see Year 1, Chapter 9, problem ZC-72on page 344 or Year 2, Chapter 8, problem GS-108 on page 319.

Example 1

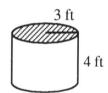

3 ft

4 ft

Find the volume of the cylinder above.

$$
\begin{aligned}
\text{Volume} &= \pi r^2 h \\
&= \pi (3)^2 (4) \\
&= 36\pi \\
&\approx 36(3.14 \\
&\approx 113.10 \text{ ft}^3
\end{aligned}
$$

Example 2

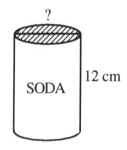

?

SODA   12 cm

The soda can above has volume 355 cm³ and height 12 cm. What is its diameter?

$$
\begin{aligned}
\text{Volume} &= \pi r^2 h \\
355 &= \pi r^2 (12) \\
\frac{355}{12\pi} &= r^2, \text{ so } \frac{355}{12(3.14)} \approx r^2 \\
9.42 &\approx r^2 \\
\text{radius} &\approx 3.06 \\
\text{diameter} &\approx 2(3.07) \approx 6.14 \text{ cm}
\end{aligned}
$$

Problems

Find the volume of each cylinder. Use 3.14 for π.

1. r = 5 cm
    h = 10 cm

2. r = 7.5 in.
    h = 8.1 in.

3. diameter = 10 cm
    h = 5 cm

4. base area = 50 cm²
    h = 4 cm

5. r = 17 cm
    h = 10 cm

6. d = 29 cm
    h = 13 cm

Find the missing part of each cylinder.

7.　　If the volume is 5175 ft³ and the height is 23 ft, find the diameter.

8.　　If the volume is 26,101.07 inches³ and the radius is 17.23 inches, find the height.

9.　　If the circumference is 126 cm and the height is 15 cm, find the volume.

Answers

1.　　785 cm³　　　　　2.　　1430.66 in³　　　　　3.　　392.50 cm³

4.　　200 cm³　　　　　5.　　9074.60 cm³　　　　　6.　　8582.41 cm³

7.　　16.93 ft　　　　　8.　　28 inches　　　　　9.　　18,960.19 cm³

VOLUME OF A CONE

Every cone has a volume that is one-third the volume of the cylinder with the same base and height. To find the volume of a cone, use the same formula as the volume of a cylinder and divide by three. The formula for the volume of a cone of base area A and height h is:

$$V = \frac{A \cdot h}{3} = \frac{\pi r^2 h}{3} \text{ or } \frac{1}{3}\pi r^2 h$$

For additional information, see Year 2, Chapter 10, problem MG-26 on page 375.

Example 1

Find the volume of the cone below.

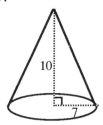

Volume $= \frac{1}{3}\pi(7)^2 \cdot 10$

$= \frac{490\pi}{3} \approx \frac{490(3.14)}{3}$

≈ 512.87 units³

Example 2

Find the volume of the cone below.

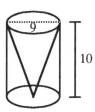

radius $= 4.5$

Volume $= \frac{1}{3}\pi(4.5)^2 \cdot 10$

$= 67.5\pi$

$\approx 67.5(3.14)$

≈ 211.95 units³

Example 3

If the volume of a cone is 4325.87 cm³ and its radius is 9 cm, find its height.

Volume $= \frac{1}{3}\pi r^2 h$

$4325.87 = \frac{1}{3}\pi(9)^2 \cdot h$

$12977.61 = \pi(81) \cdot h$

$\frac{12977.61}{81\pi} = h$

$\frac{12977.61}{81(3.14)} \approx h$

51.02 cm $\approx h$

Problems

Find the volume of each cone. Use 3.14 for π.

1. $r = 4$ cm
   $h = 10$ cm

2. $r = 2.5$ in.
   $h = 10.4$ in.

3. $d = 12$ in.
   $h = 6$ in.

4. $d = 9$ cm
   $h = 10$ cm

5. $r = 6\frac{1}{3}$ ft
   $h = 12\frac{1}{2}$ ft

6. $r = 3\frac{1}{4}$ ft
   $h = 6$ ft

Find the missing part of each cone described below.

7. If $V = 1000$ cm^3 and $r = 10$ cm, find h.

8. If $V = 2000$ cm^3 and $h = 15$ cm, find r.

9. If the circumference of the base $= 126$ cm and $h = 10$ cm, find the volume.

Answers

1. 167.47 cm^3

2. 68.03 in.^{3}

3. 226.08 in.^{3}

4. 211.95 cm^3

5. 524.79 ft^3

6. 66.33 ft^3

7. 9.55 cm

8. 11.29 cm

9. 4211.38 cm^3

THE PYTHAGOREAN THEOREM

A right triangle is a triangle in which the two shorter sides form a right angle. The shorter sides are called legs. Opposite the right angle is the third and longest side called the hypotenuse.

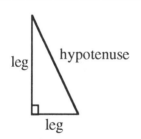

The Pythagorean Theorem states that for any right triangle, the sum of the squares of the lengths of the legs is equal to the square of the length of the hypotenuse.

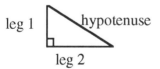

$$(\textbf{leg 1})^2 + (\textbf{leg 2})^2 = (\textbf{hypotenuse})^2$$

For additional information, see Year 2, Chapter 8, problems GS-1 through 6 on pages 291-93.

Example 1

Use the Pythagorean Theorem to find x.

a)

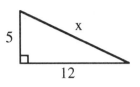

$$5^2 + 12^2 = x^2$$
$$25 + 144 = x^2$$
$$169 = x^2$$
$$13 = x$$

b)

$$x^2 + 8^2 = 10^2$$
$$x^2 + 64 = 100$$
$$x^2 = 36$$
$$x = 6$$

Example 2

Not all problems will have exact answers. Use square root notation and your calculator. Round to the nearest hundredth.

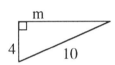

$$4^2 + m^2 = 10^2$$
$$16 + m^2 = 100$$
$$m^2 = 84$$
$$m = \sqrt{84} \approx 9.17$$

Example 3

A guy wire is needed to support a tower. The wire is attached to the ground five meters from the base of the tower. How long is the wire if the tower is 10 meters tall?

First draw a diagram to model the problem, then write an equation using the Pythagorean Theorem and solve it.

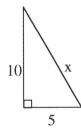

$$x^2 = 10^2 + 5^2$$
$$x^2 = 100 + 25$$
$$x^2 = 125$$
$$x = \sqrt{125} \approx 11.18 \text{ cm}$$

Problems

Write an equation and solve it to find the length of the unknown side. Round answers to the nearest hundredth.

1.
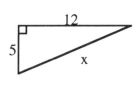

2.
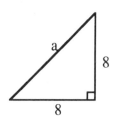

3.
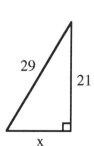

4.
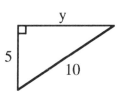

5.
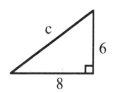

6.
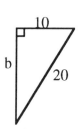

Draw a diagram, write an equation, and solve it. Round answers to nearest hundredth.

7.   Find the diagonal of a television screen 30 inches wide by 35 inches tall.

8.   A four meter ladder is one meter from the base of a building. How high up the building will the ladder reach?

9.   Sam drove eight miles south and then five miles west. How far is he from his starting point?

10. The length of the hypotenuse of a right triangle is six centimeters. If one leg is four centimeters, how long is the other leg?

11. Find the length of a path that runs diagonally across a 55 yard by 100 yard field.

12. How long an umbrella will fit in the bottom of a suitcase 1.5 feet by 2.5 feet?

Answers

1.   13   2.   11.31   3.   20   4.   8.66   5.   10   6.   17.32
7.   46.10 in. 8.   3.87 in. 9.   9.43 mi   10. 4.47 cm   11. 114.13 yd 12. 2.92 ft

ANGLES, TRIANGLES, AND QUADRILATERALS

PROPERTIES OF ANGLE PAIRS

Angles are measured in degrees (°). A right angle is 90° (clock hands at 3 o'clock). Angles measuring between 0° and 90° are called acute, angles measuring between 90° and 180° are called obtuse.

Intersecting lines form four angles. The pairs of angles across from each other are called vertical angles. The measures of vertical angles are equal.

$\angle$ x and $\angle$ y are vertical angles
$\angle$ w and $\angle$ z are vertical angles

If the sum of the measures of two angles is exactly 180°, then they are called supplementary angles.

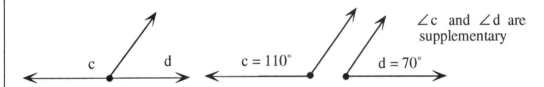

$\angle$ c and $\angle$ d are supplementary

If the sum of the measures of two angles is exactly 90°, then they are called complementary angles.

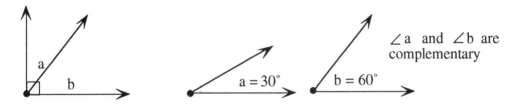

$\angle$ a and $\angle$ b are complementary

Angles that share a vertex and one side but have no common interior points (that is, do not overlap each other) are called adjacent angles.

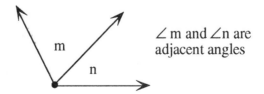

$\angle$ m and $\angle$ n are adjacent angles

For additional information, see Year 1, Chapter 8, problems MC-2 on page 274, problems MC-22 through 26 on pages 281-84, and problem MC-42 on page 288.

Example 1

Find the measure of the missing angles if m $\angle 3 = 50°$

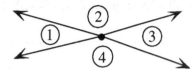

- m $\angle 1$ = m $\angle 3$ (vertical angles)
  $\Rightarrow$ m $\angle 1 = 50°$
- $\angle 2$ and $\angle 3$ (supplementary angles)
  $\Rightarrow$ m $\angle 2 = 180° - 50° = 130°$
- m $\angle 2$ = m $\angle 4$ (vertical angles)
  $\Rightarrow$ m $\angle 4 = 130°$

Example 2

Classify each pair of angles below as vertical, supplementary, complementary, or adjacent.

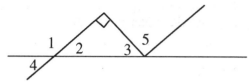

a) $\angle 1$ and $\angle 2$ are adjacent and supplementary
b) $\angle 2$ and $\angle 3$ are complementary
c) $\angle 3$ and $\angle 5$ are adjacent
d) $\angle 1$ and $\angle 4$ are adjacent and supplementary
e) $\angle 2$ and $\angle 4$ are vertical

Problems

Find the measure of each angle labeled with a variable by calculation. Do not measure them.

1.

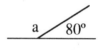

2.

3.

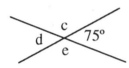

4.

5.

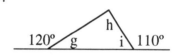

6.

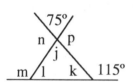

Answers

1. m $\angle a = 100°$

2. m $\angle b = 55°$

3. m $\angle c = 105°$
    m $\angle d = 75°$
    m $\angle e = 105°$

4. m $\angle f = 50°$

5. m $\angle g = 60°$
    m $\angle h = 50°$
    m $\angle i = 70°$

6. m $\angle j = 75°$ m $\angle k = 65°$
    m $\angle l = 40°$ m $\angle m = 140°$
    m $\angle n = 105°$ m $\angle p = 105°$

ISOSCELES AND EQUILATERAL TRIANGLES AND EQUATIONS IN GEOMETRIC CONTEXT

For additional information, see Year 1, Chapter 8, problems MC-53 on page 291, MC-58 on page 296, and MC-86on page 301.

Examples

a) Find the measure of $\angle$ x.

b) Solve for x.

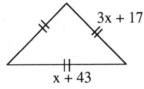

c) Solve for x.

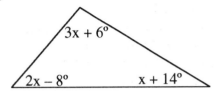

Since we have an isosceles triangle, both base angles measure 70°. The sum of the measures of the angles of a triangle is 180° so:

$$x + 70° + 70° = 180°$$
$$x + 140° = 180°$$
$$x = 40°$$

Since we have an equilateral triangle, all sides are of equal length so:

$$3x + 17 = x + 43$$
$$2x + 17 = 43$$
$$2x = 26$$
$$x = 13$$

The sum of the measures of the angles of a triangle is 180° so:

$$(2x - 8°) + (3x + 6°) + (x + 14°) = 180°$$
$$6x + 12° = 180°$$
$$6x = 168°$$
$$x = 28°$$

Problems

In each problem, solve for the variable.

1.

2.

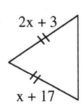

3.

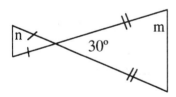

4.

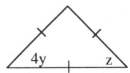

5.

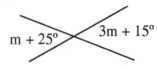

6.

7.

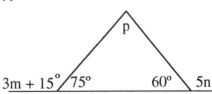

8.

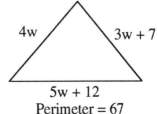

5w + 12
Perimeter = 67

9.

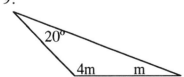

127

Answers

1. $a = 69°$
2. $x = 14$
3. $m = 75°, \ n = 75°$
4. $y = 15°, \ z = 60°$
5. $m = 5°$
6. $x = 63°, \ y = 54°$
7. $m = 30°, \ n = 24°, \ p = 45°$
8. $w = 4$
9. $m = 32°$

QUADRILATERALS

A quadrilateral is any four-sided shape. There are five special cases of quadrilaterals with which students should be familiar.

*   Trapezoid – a quadrilateral with one pair of parallel sides.

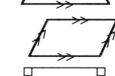

*   Parallelogram – a quadrilateral with two pairs of parallel sides.

*   Rectangle – a quadrilateral with four right angles.

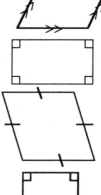

*   Rhombus – a quadrilateral with four congruent sides.

*   Square – a quadrilateral with four right angles **and** four congruent sides.

In Year 1, Chapter 8, problems MC-55, 56, and MC-57 on pages 292-95 students learn that the sum of the measures of the angles in a quadrilateral equals 360°.

Example

Find the measure of $\angle x$.

$$x + 128° + 95° + 58° = 360°$$
$$x + 281° = 360°$$
$$x = 79°$$

Problems

In each problem solve for the variable.

1.

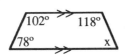

2.

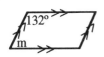

3.

4.

5.

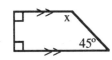

6.

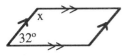

Answers

1. 62°   2. 48°   3. 90°   4. 57°   5. 135°   6. 148°

SCALE FACTOR AND RATIOS OF GROWTH

RATIOS OF DIMENSIONAL CHANGE FOR TWO-DIMENSIONAL FIGURES

Geometric figures can be reduced or enlarged. When this change happens, every length of the figure is reduced or enlarged equally. For additional information, see Year 2, Chapter 6, problems RS-72, 73, and 75 on pages 227-29 and RS-88 on page 233 for examples.

The ratio of any two corresponding sides of the original and new figure is called a scale factor. In this book we always place new figure measurements over their original figure measurements in a scale ratio.

Example

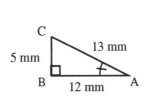

original triangle

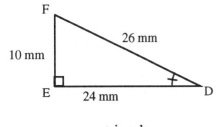

new triangle

Side length ratios:

$$\frac{DE}{AB} = \frac{24}{12} = \frac{2}{1}$$

$$\frac{FD}{CA} = \frac{26}{13} = \frac{2}{1}$$

$$\frac{FE}{CB} = \frac{10}{5} = \frac{2}{1}$$

The scale factor for length is 2 to 1.

The scale factor for area is 4 to 1.

Perimeter ratio:

$$\frac{\Delta DEF}{\Delta ABC} = \frac{60}{30} = \frac{2}{1}$$

Area ratio:

$$\frac{\Delta DEF}{\Delta ABC} = \frac{120}{30} = \frac{4}{1} \text{ or } \left(\frac{2}{1}\right)^2$$

Problems

Determine the scale factor for each pair of figures below.

1.

Original   New

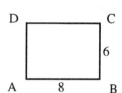

2.

Original   New

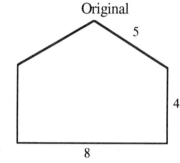
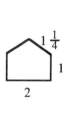

Find the perimeter and the area scale factor for each pair of figures below.

3.

Original   New

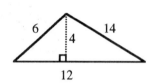

4.

Original   New

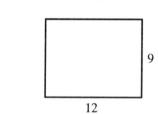

5. Two triangles are similar. The new triangle was enlarged by a scale factor of $\frac{3}{1}$. Use a proportion to solve the problems below.

    a) If the original triangle has a perimeter of 12, what is the perimeter of the new triangle?

    b) If the original triangle has an area of 6, what is the area of the new triangle?

6. Two rectangles are similar. The new rectangle was reduced by a scale factor of $\frac{1}{4}$. Use a proportion to solve the problems below.

    a) If the original rectangle has a perimeter of 24, what is the perimeter of the new rectangle?

    b) If the original area is 32 square units, what is the new area?

Answers

1. $\dfrac{4}{8} = \dfrac{1}{2}$

2. $\dfrac{2}{8} = \dfrac{1}{4}$

3. perimeter scale factor = $\dfrac{32}{16} = \dfrac{2}{1}$

    area scale factor = $\dfrac{24}{6} = \dfrac{4}{1}$

4. perimeter scale factor = $\dfrac{14}{42} = \dfrac{1}{3}$

    area scale factor = $\dfrac{12}{108} = \dfrac{1}{9}$

5. a) 36 units   b) 54 sq. units

6. a) 6 units   b) 2 sq. units

RATIOS OF DIMENSIONAL CHANGE FOR THREE-DIMENSIONAL OBJECTS

When looking at three-dimensional shapes and enlarging or reducing them, a third dimension (height) is taken into consideration. The volume scale factor then becomes the cube of the scale factor (length ratio).

Volume of the original = $2 \cdot 4 \cdot 1 = 8$ units^3

Volume of the enlargement = $6 \cdot 12 \cdot 3 = 216$ units^3

Scale factor of volume = $\dfrac{216}{8} = \dfrac{27}{1}$.

Original

By enlarging each edge by a scale factor of $\dfrac{3}{1}$, the new volume has a scale factor of $\left(\dfrac{3}{1}\right)^3 = \dfrac{3^3}{1^3} = \dfrac{27}{1}$.

For additional information, see Year 2, Chapter 10, problem MG-14 on page 371.

Enlargement

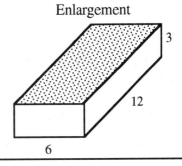

Example

Suppose you want to find the surface area and volume of a prism with dimensions three times as large as the figure at right. You could triple the dimensions and do the calculations OR use ratios. Here we use proportions to find the surface area and volume of the new prism after calculating them for the original figure.

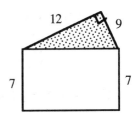

a)   Calculate the surface area and volume of the figure at right.

$B = \dfrac{12 \cdot 9}{2} = 54$ units^2, so $V = 54 \cdot 7 = 378$ units^3

$SA = 2(54) + 12 \cdot 7 + 9 \cdot 7 + 15 \cdot 7$ (Note: 15 comes from $\sqrt{12^2 + 9^2} = 15$.)
   $= 108 + 252$
   $= 360$ units^2

b) Use ratios to find the enlarged figure's surface area and volume.

SA: length ratio = $\frac{3}{1}$ $\Rightarrow$ area = $\left(\frac{3}{1}\right)^2 = \frac{9}{1}$, so $\frac{9}{1} = \frac{SA}{360}$ $\Rightarrow$ SA = 3240 units²

V: volume ratio = $\left(\frac{3}{1}\right)^3 = \frac{27}{1}$ $\Rightarrow$ $\frac{27}{1} = \frac{V}{378}$ $\Rightarrow$ V = 10,206 units³

SUMMARY: RATIOS OF DIMENSIONAL CHANGE

For any pair of similar figures or solids with a scale factor $\frac{a}{b}$, the enlargement and reduction relationships are:

Length (one dimension)	Area (two dimensions)	Volume (three dimensions)
$\frac{a}{b}$	$\frac{a^2}{b^2}$	$\frac{a^3}{b^3}$

Length ratios apply to sides, edges, and perimeters of figures.

Area ratios apply to surface area of solids and the area of two-dimensional regions.

See Year 2, Chapter 10, problem MG-14 on page 371 for the complete Tool Kit entry about dimensional change.

Problems

1. Reduce the prism at right by a scale factor of $\frac{1}{2}$. Find the original volume and compare it to the new volume. That is, determine $\frac{\text{new volume}}{\text{old volume}}$.

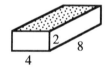

2. The dimensions of a cube are 4 by 4 by 4. Suppose the cube is enlarged by a scale factor of $\frac{2}{1}$. Compare the volumes.

3. If you have a square prism shown at right and you enlarge it by a scale of 5 to 1, what is the new volume?

4. If you reduce a prism with a volume of 81 units³ by a scale factor of $\frac{1}{3}$, what is the new volume?

5. A prism with an original volume of 24 inches³ is enlarged. The resulting new volume is 192 inches³. What is the scale factor used for enlarging?

Answers

1. The new dimensions are 2 by 4 by 1, so V = 8 units^3.

   $$\frac{8}{64} = \frac{1}{8}$$

2. The new dimensions are 8 by 8 by 8, so V = 512.

   $$\frac{512}{64} = \frac{8}{1}$$

3. Volume = $4 \cdot 4 \cdot 3 = $ 48 units^3.

   Scale factor for volume is $\left(\frac{5}{1}\right)^3 = \frac{125}{1}$.

   $$\frac{125}{1} = \frac{V}{48}; \ V = 6000 \text{ units}^3$$

4. $\frac{3}{81} = \frac{1}{27}$; new volume = 3 units^3

5. $\frac{192}{24} = \frac{8}{1}$; scale factor 2 : 1

Statistics, Data Analysis, and Probability

Measures of Central Tendency

Sampling Populations

Correlation

Probability

Tree Diagrams

MEASURES OF CENTRAL TENDENCY

The measures of central tendency are numbers that locate or approximate the "center" of a set of data. Mean, median, and mode are the most common measures of central tendency.

The mean is the arithmetic average of a data set. Add all the values in a set and divide this sum by the number of values in the set.

For additional information, see Year 1, Chapter 1, problems AR-26 and 27 on page 11 and AR-37 on page 15 or Year 2, Chapter 1, problems GO-30 on page 14 and GO-64 and 65 on pages 26-27.

Example 1

Find the mean of this set of data: 34, 31, 37, 44, 38, 34, 42, 34, 43, and 41.

- $34 + 31 + 37 + 44 + 38 + 34 + 42 + 34 + 43 + 41 = 378$

- $378 \div 10 = 37.8$

The mean of this set of data is 37.8.

Example 2

Find the mean of this set of data: 92, 82, 80, 92, 78, 75, 95, and 77.

- $92 + 82 + 80 + 92 + 78 + 75 + 95 + 77 + 77 = 748$

- $748 \div 9 = 83.11$

The mean of this set of data is 83.11.

Problems

Find the mean of each set of data.

1. 29, 28, 34, 30, 33, 26, and 34.

2. 25, 34, 35, 27, 31, and 30.

3. 80, 89, 79, 84, 95, 79, 78, 89, 76, 82, 76, 92, 89, 81, and 123.

4. 116, 104, 101, 111, 100, 107, 113, 118, 113, 101, 108, 109, 105, 103, and 91.

The mode is the value in a data set that occurs most often. Data sets may have more than one mode.

Example 3

Find the mode of this set of data: 34, 31, 37, 44, 34, 42, 34, 43, and 41.

- The mode of this data set is 34 since there are three 34s and only one of each of the other numbers.

Example 4

Find the mode of this set of data: 92, 82, 80, 92, 78, 75, 95, 77, and 77.

- The modes of this set of data are 77 and 92 since there are two of each of these numbers and only one of each of the other numbers. This data set is said to be bimodal since it has two modes.

Statistics, Data Analysis, and Probability

Problems

Find the mode of each set of data.

5. 29, 28, 34, 30, 33, 26, and 34.

6. 25, 34, 35, 27, 25, 31, and 30.

7. 80, 89, 79, 84, 95, 79, 89, 76, 82, 76, 92, 89, 81, and 123.

8. 116, 104, 101, 111, 100, 107, 113, 118, 113, 101, 108, 109, 105, 103, and 91.

The median is the middle number in a set of data <u>arranged</u> <u>in</u> <u>numerical</u> <u>order</u>. If there are an even number of values, the median is the mean of the two middle numbers.

Example 5

Find the median of this set of data: 34, 31, 37, 44, 38, 34, 43, and 41.

- Arrange the data in order: 31, 34, 34, 37, 38, 41, 43, 44.

- Find the middle value(s): 37 and 38.

- Since there are two middle values, find their mean: $37 + 38 = 75$, $75 \div 2 = 37.5$. Therefore, the median of this data set is 37.5.

Example 6

Find the median of this set of data: 92, 82, 80, 92, 78, 75, 95, 77, and 77.

- Arrange the data in order: 75, 77, 77, 78, 80, 82, 92, 92, and 95.

- Find the middle value(s): 80. Therefore, the median of this data set is 80.

Problems

Find median of each set of data.

9. 29, 28, 34, 30, 33, 26, and 34.

10. 25, 34, 27, 25, 31, and 30.

11. 80, 89, 79, 84, 95, 79, 78, 89, 76, 82, 76, 92, 89, 81, and 123.

12. 116, 104, 101, 111, 100, 107, 113, 118, 113, 101, 108, 109, 105, 103, and 91.

The range of a set of data is the difference between the highest value and the lowest value.

Example 7

Find the range of this set of data: 114, 109, 131, 96, 140, and 128.

- The highest value is 140.

- The lowest value is 96.

- $140 - 96 = 44$.

- The range of this set of data is 44.

Example 8

Find the range of this set of data: 37, 44, 36, 29, 78, 15, 57, 54, 63, 27, and 48.

- The highest value is 78.

- The lowest value is 15.

- $78 - 15 = 63$.

- The range of this set of data is 63.

Problems

Find the range of each set of data in problems 9 through 12.

Outliers are numbers in a data set that are either much higher or much lower that the other numbers in the set.

Example 9

Find the outlier of this set of data: 88, 90 96, 93, 87, 12, 85, and 94.

- The outlier is 12.

Example 10

Find the outlier of this set of data: 67, 54, 49, 76, 64, 59, 60, 72, 123, 44, and 66.

- The outlier is 123.

Problems

Find the outlier for each set of data.

13. 70, 77, 75, 68, 98, 70, 72, and 71.

14. 14, 22, 17, 61, 20, 16, and 15.

15. 1376, 1645, 1783, 1455, 3754, 1790, 1384, 1643, 1492, and 1776.

16. 62, 65, 93, 51, 55, 14, 79, 85, 55, 72, 78, 83, 91, and 76.

A stem-and-leaf plot is a way to display data that shows the individual values from a set of data and how the values are distributed. This type of display clearly shows median, mode, range, and outliers. The "stem" part on the graph represents the leading digit(s) of the number. The "leaf" part of the graph represents the other digit(s).

For additional information, see Year 1, Chapter 1, problems AR-38, 39, and 44 on pages 15-17 or Year 2, Chapter 1, problem GO-45 on page 19.

Example 12

Make a stem-and-leaf plot of this set of data: 34, 31, 37, 44, 38, 29, 34, 42, 43, 34, 52, and 41.

```
2 | 9
3 | 1 4 4 4 7 8
4 | 1 2 3 4
5 | 2
```

Example 13

Make a stem-and-leaf plot of this set of data: 92, 82, 80, 92, 78, 75, 95, 77, and 77.

```
7 | 5 7 7 8
8 | 0 2
9 | 2 2 5
```

Problems

Make a stem-and-leaf plot of each set of data.

17. 29, 28, 34, 30, 33, 26, 18, and 34.

18. 25, 34, 27, 25, 19, 31, 42, and 30.

19. 80, 89, 79, 84, 95, 79, 89, 67, 82, 76, 92, 89, 81, and 123.

20. 116, 104, 101, 111, 100, 107, 113, 118, 113, 101, 108, 109, 105, 103, and 91.

A way to display data that shows how the data is grouped or clustered is a box-and-whisker plot. The box-and-whisker plot displays the data using quartiles. This type of display clearly shows the median, range, and outliers of a data set. It is a very useful display for comparing sets of data.

For additional information, see Year 2, Chapter 1, problem GO-65 on page 27.

Example 14

Display this data in a box-and-whisker plot: 51, 55, 55, 62, 65, 72, 76, 78, 79, 82, 83, 85, 91, and 93.

- Since this data is already in order from least to greatest, it can be seen that the range is $93 - 51 = 42$. Thus you start with a number line with equal intervals from 50 to 95.

- The median of the set of data is 77. A line is drawn at this value above the number line.

- The median of the lower half of the data (the lower quartile) is 62. A line is drawn at this value above the number line.

- The median of the upper half of the data (the upper quartile) is 83. A line is drawn at this value above the number line.

- A box is drawn between the upper and lower quartiles.

- Place a dot at the minimum value (51) and a dot at the maximum value (93). The lines which connect these dots to the box are called the whiskers.

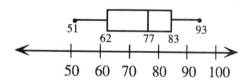

Example 15

Display this data in a box-and-whisker plot: 62, 65, 93, 51, 12, 79, 85, 55, 72, 78, 83, 91, and 76.

- Place the data in order from least to greatest: 12, 51, 55, 62, 65, 72, 76, 78, 79, 83, 85, 91, 93. The range is $93 - 12 = 81$. Thus you want a number line with equal intervals from 10 to 100.

- Find the median of the set of data: 76. Draw the line.

- Find the lower quartile: $55 + 62 = 117$; $117 \div 2 = 58.5$. Draw the line.

- Find the upper quartile: $83 + 85 = 168$; $168 \div 2 = 84$. Draw the line.

- Draw the box connecting the upper and lower quartiles. Place a dot at the minimum value (12) and a dot at the maximum value (93). Draw the whiskers.

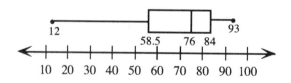

Problems

Create a stem-and-leaf plot and a box-and-whisker plot for each set of data in problems 21 through 24.

21.   45, 47, 52, 85, 46, 32, 83, 80, and 75.

22.   75, 62, 56, 80, 72, 55, 54, and 80.

23.   49, 54, 52, 58, 61, 72, 73, 78, 73, 82, 83, 73, 61, 67, and 68.

24.   65, 35, 48, 29, 57, 87, 94, 68, 86, 73, 58, 74, 85, 91, 88, and 97.

25.   Given a set of data: 265, 263, 269, 259, 267, 264, 253, 275, 264, 260, 273, 257, and 291.

    a)   Make a stem-and-leaf plot of this data.

    b)   Find the mean, median, and mode of this data.

    c)   Find the range of this data.

    d)   Make a box-and-whisker plot for this data.

26.   Given a set of data: 48, 42, 37, 29, 49, 46, 38, 28, 45, 45, 35, 46.25, 34, 46, 46.5, 43, 46.5, 48, 41.25, 29, and 47.75.

    a)   Make a stem-and-leaf plot of this data.

    b)   Find the mean, median, and mode of this data.

    c)   Find the range of the data.

    d)   Make a box-and-whisker plot for this data.

Answers

1.   30.57

2.   30.$\overline{3}$

3.   86.1$\overline{3}$

4.   106.8$\overline{6}$

5.   34

6.   25

7.   89

8.   101 and 113

9.   median 30; range 8

10.   median 28.5; range 9

11.   median 82; range 47

12.   median 107; range 27

13.   98

14.   61

15.   3754

16.   14

17.

```
1 | 8
2 | 6 8 9
3 | 0 3 4 4
```

18.

```
1 | 9
2 | 5 5 7
3 | 0 1 4
4 | 2
```

19.

```
 6 | 7
 7 | 6 9 9
 8 | 0 1 2 4 9 9 9
 9 | 2 5
10 |
11 |
12 | 3
```

20.

```
 9 | 1
10 | 0 1 1 3 4 5 7 8 9
11 | 1 3 3 6 8
```

21.

3	2
4	5 6 7
5	2
6	
7	5
8	0 3 5

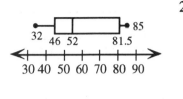

22.

5	4 5 6
6	2
7	2 5
8	0 0

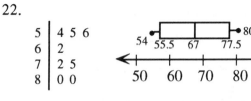

23.

4	9
5	2 4 8
6	1 1 7 8
7	2 3 3 3 8
8	2 3

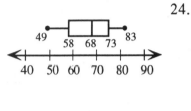

24.

2	9
3	5
4	8
5	7 8
6	5 8
7	3 4
8	5 6 7 8
9	1 4 7

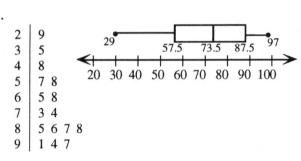

25.

25	3 7 9
26	0 3 4 4 5 7 9
27	3 5
28	
29	1

Mean: 266.15
Median: 264
Mode: 264
Range: 38

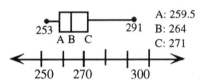

A: 259.5
B: 264
C: 271

26.

2	8 9 9
3	4 5 7 8
4	1.25 2 3 5 5 6 6 6.25 6.5 6.5 7.75 8 8 9

Mean: 39.56
Median: 45
Mode: 29, 45, 46.5, 46, and 48
Range: 21

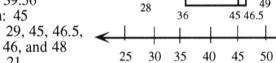

SAMPLING POPULATIONS

In order to have students conduct a representative survey, they are introduced to different types of surveys and methods for selecting survey samples. The vocabulary is introduced in Year 1, Chapter 10, problem JM-90 on page 387.

A population is a collection of objects or a group of people about whom information is gathered.

A sample is a subgroup of the population. For example, if you want to conduct a survey at your school about what foods to serve in the cafeteria, the population would be the entire student body.

A representative sample is a subgroup of the population that matches the general characteristics of the entire population. If you choose to sample 10% of the students, you need to include the correct fraction of students from each grade and an equal number of male and female students.

Problems

Answer the following questions.

1. If survey results were published in an advertisement that 3 out of 4 people surveyed said they preferred the advertised product, what questions would you ask about the people surveyed?

2. What can people making surveys do to ensure that the survey is truly random?

Answers

1. How many people were surveyed? Who did the people surveyed represent? Could all the people surveyed be employees of the advertiser? Who conducted the survey? What questions were asked?

2. Determine the general characteristics of the total population being surveyed and make sure a subgroup representing each characteristic is surveyed.

CORRELATION

A scatter plot like the first one at right, appears to have a positive correlation, since people seem to own more cars as they increase in age. The second scatter plot shows a negative correlation, where the number of incorrect items on tests decreases as more tests are taken.

Students should begin to understand the concept that a positive correlation exists if the items on both axes tend to increase and that a negative correlation exists if the items on the y-axis decrease as the items on the x-axis increase. The closer the points come to approaching a straight line, the stronger the correlation.

One caution: It is easy to jump to the conclusion that if there is a strong correlation, one factor causes the other. This is not necessarily true. For example, a scatter plot showing the relationship of height to reading ability might show a very strong correlation but neither increased height increases reading ability, nor does increased reading ability cause growth. In this case both height and reading ability are age related.

For additional information, see Year 1, Chapter 10, problems JM-116 through 119 on pages 395-97 or see Year 2, Chapter 1, problem GO-71 on pages 29-30 and problems GO-84 through 86 on pages 33-34.

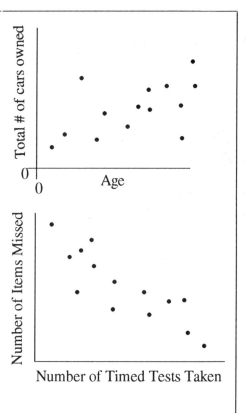

Problems

Use the scatter plot at right to answer the following questions:

1. Is there a correlation between age and height in girls?

2. If there is a correlation, is it positive or negative?

3. What does your answer to problem 2 mean?

4. What will happen to the scatter plot if the x-axis is increased to age 40?

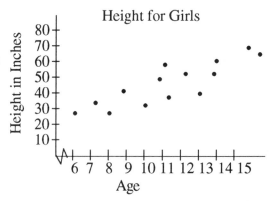

Answers

1. yes   2. positive   3. Girls increase in height as they age.

4. The correlation would continue until it leveled off somewhere between ages 15 and 20. From that point, height should stay the same since growth has ended.

Probability: A number between zero and one that states the likelihood of an event occurring. It is the ratio of specified outcomes to all possible outcomes (the sample space).

Outcome: Any possible or actual result or consequence of the action(s) considered, such as rolling a five on a die or getting tails when flipping a coin.

Event: An outcome or group of outcomes from an experiment, such as rolling an even number on a die.

Probability is the likelihood that a specific outcome will occur. If all the outcomes of an event are equally likely to occur, then the probability that a specified outcome occurs is:

$$P(\text{outcome}) = \frac{\text{number of ways that the specified outcome occurs}}{\text{total number of possible outcomes}}$$

Fractions are used to express the probability that certain events will (or will not) happen. The denominator of the fraction shows the total number of outcomes for an event. The numerator of the fraction indicates the number of times an event could happen.

$$\text{Theoretical probability} = \frac{\text{number of event outcomes}}{\text{total number of possible outcomes}}$$

Experimental probability refers to the occurrence of an event when the activity is actually done.

Two events are dependent if the outcome of the first event affects the outcome of the second event. For example, if you draw a card from a deck and do not replace it for the next draw, the two events – drawing one card without replacing it, then drawing a second card – are dependent.

Two events are independent if the outcome of the first event does not affect the outcome of the second event. For example, if you draw a card from a deck but replace it before you draw again, the two events are independent.

For additional information, see Year 1, Chapter 10, problems JM-7, 13, and 17 on pages 363-66 or Year 2, Chapter 3, problem MD-5 on page 85.

Example 1

If you roll a fair, 6-sided die, what is P(3), that is, the probability that you will roll a 3?

Because the six sides are equally likely to come up, and there is only one 3, $P(3) = \frac{1}{6}$.

Example 2

There are 5 marbles in a bag: 2 clear, 1 green, 1 yellow, and 1 blue.
If one marble is chosen randomly from the bag, what is the probability that it will be yellow?

$$P(\text{yellow}) = \frac{1 \text{ (yellow)}}{5 \text{ (outcomes)}} = \frac{1}{5}$$

Problems

Answer these questions.

1. There are six crayons in a box: 1 black, 1 white, 1 red, 1 yellow, 1 blue, and 1 green. What is the probability of randomly choosing a green?

2. A spinner is divided into four equal sections numbered 2, 4, 6, and 8. What is the probability of spinning an 8?

3. A fair die numbered 1, 2, 3, 4, 5, and 6 is rolled. What is the chance that an even number will be rolled?

4. The light is out in Sara's closet, so she cannot see the clothes that are hanging there. She has three t-shirts hanging in front of her: 1 brown, 1 black, and 1 navy blue. What is the probability that she chooses the black one?

Answers

1. $\frac{1}{6}$
2. $\frac{1}{4}$
3. $\frac{3}{6}$ or $\frac{1}{2}$
4. $\frac{1}{3}$

Example 3

Joe flipped a coin 50 times. When he recorded his tosses, his result was 30 heads and 20 tails. Joe's activity provided data to calculate experimental probability for flipping a coin.

a) What is the theoretical probability of Joe flipping heads?

The theoretical probability is 50% or $\frac{1}{2}$, because there are only two possibilities (heads and tails), and each is equally likely to occur.

b) What was the experimental probability of flipping a coin and getting heads based on Joe's activity?

The experimental probability is $\frac{30}{50}$, $\frac{3}{5}$, or 60%. These are the results Joe actually got when he flipped the coin.

c) Are these dependent or independent events?

These are independent events. Each flip still has the same possible outcomes (heads or tails), and you can only get one result at a time.

Example 4

Decide whether these statements are theoretical or experimental.

a) The chance of rolling a 6 on a fair die is $\frac{1}{6}$.
This statement is theoretical.

b) I rolled the die 12 times and 5 came up three times.
This statement is experimental.

c) There are 15 marbles in a bag; 5 blue, 6 yellow, and 4 green. The probability of getting a blue marble is $\frac{1}{3}$. This statement is theoretical.

d) When Veronika pulled three marbles out of the bag she got 2 yellow and 1 blue, or $\frac{2}{3}$ yellow, $\frac{1}{3}$ blue. This statement is experimental.

Example 5

Juan pulled a red card from the deck of regular playing cards. This probability is $\frac{26}{52}$ or $\frac{1}{2}$. He puts the card back into the deck. Will his chance of pulling a red card next time change?

No, his chance of pulling a red card next time will not change, because he replaced the card. There are still 26 red cards out of 52. This is an example of an independent event; his pulling out and replacing a red card does not affect any subsequent selections from the deck.

Example 6

Brett has a bag of 30 multi-colored candies. 15 are red, 6 are blue, 5 are green, 2 are yellow, and 2 are brown. If he pulls out a yellow candy and eats it, does this change his probability of pulling any other candy from the bag?

Yes, this changes the probability, because he now has only 29 candies in the bag and only 1 yellow candy. Originally, his probability of yellow was $\frac{2}{30}$ or $\frac{1}{15}$; it is now $\frac{1}{29}$. Similarly, red was $\frac{15}{30}$ or $\frac{1}{2}$ and now is $\frac{15}{29}$, better than $\frac{1}{2}$. This is an example of a dependent event.

Problems

Decide whether these are independent or dependent events.

1. Flipping two coins.

2. Taking a black 7 out of a deck of cards and not returning it.

3. Taking a red licorice from a bag and eating it.

Answers

1. independent   2. dependent   3. dependent

PROBABILITY FOR TWO OR MORE EVENTS

Addition and multiplication are used to determine the likelihood of an event occurring in more complex situations.

Addition of fractions is used in the following example when we are trying to determine the probability of two or more events.

Example 7

A spinner is divided into five equal sections numbered 1, 2, 3, 4, and 5. What is the probability of spinning a 2 or a 5?

Step 1: Determine both probabilities: $P(2) = \frac{1}{5}$ and $P(5) = \frac{1}{5}$

Step 2: Add the fractions describing each probability: $\frac{1}{5} + \frac{1}{5} = \frac{2}{5}$

The probability of spinning a 2 or a 5 is $\frac{2}{5}$: $P(2 \text{ or } 5) = \frac{2}{5}$

Problems

Answer the following questions.

1. One die, numbered 1, 2, 3, 4, 5, and 6, is rolled. What is the probability of rolling a 1 or a 6?

2. A spinner is divided into eight equal sections. The sections are numbered 1, 2, 3, 4, 5, 6, 7, and 8. What is the probability of spinning a 2, 3, or a 4?

3. Patty has a box of 12 colored pencils. There are 2 blue, 1 black, 1 gray, 3 red, 2 green, 1 orange, 1 purple, and 1 yellow. Patty closes her eyes and chooses one pencil. She is hoping to choose a green or a red. What is the probability she will get her wish?

4. John has a bag of jelly beans. There are 100 beans in the bag. $\frac{1}{4}$ of the beans are cherry, $\frac{1}{4}$ of the beans are orange, $\frac{1}{4}$ of the beans are licorice, and $\frac{1}{4}$ of the beans are lemon. What is the probability that John will chose one of his favorite flavors, orange or cherry?

Answers

1. $\frac{2}{6}$ or $\frac{1}{3}$ 2. $\frac{3}{8}$ 3. $\frac{5}{12}$ 4. $\frac{2}{4}$ or $\frac{1}{2}$

Multiplication of fractions is used in the following example where the desired outcome is the product of two possibilities.

Example 8

If each of the regions in each spinner at right is the same size, what is the probability of spinning each spinner and getting a green t-shirt?

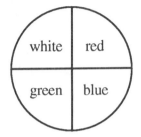

Step 1: Determine both possibilities:

$P(green) = \frac{1}{4}$ and $P(t\text{-shirt}) = \frac{1}{3}$

Step 2: Multiply both probabilities: $\frac{1}{4} \cdot \frac{1}{3} = \frac{1}{12}$

The probability of spinning a green t-shirt is $\frac{1}{12}$: $P(green\ t\text{-shirt}) = \frac{1}{12}$

You can also use a probability rectangle to organize the information from this problem. For more information, refer to Year 2, Chapter 3, problems MD-15 and 16 on pages 88-89.

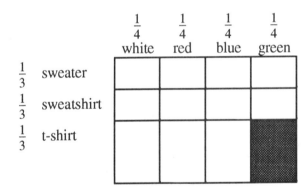

Each box in the rectangle represents a combination of a color and a sweater, sweatshirt, or a t-shirt. The area of each box represents the probability of getting each combination. The shaded region represents the probability of getting a green t-shirt: $\frac{1}{4} \cdot \frac{1}{3} = \frac{1}{12}$.

Problems

Determine the following probabilities. A probability rectangle may be useful in some of the problems below.

1. If each section in each spinner is the same size, what is the probability of getting a blue bicycle?

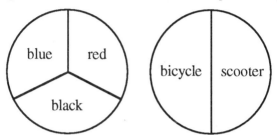

2. What is the probability of spinning each spinner below and getting a red pencil?

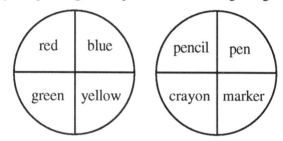

3. Mary is playing a game in which she rolls one die and spins a spinner. What is the probability she will get the 3 and black she needs to win the game?

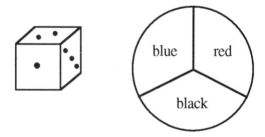

4. Use the spinners below to tell Paul what his chances are of getting the silver truck he wants.

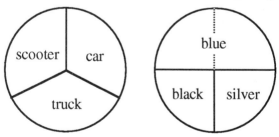

Answers

1. $\frac{1}{6}$ 2. $\frac{1}{16}$ 3. $\frac{1}{18}$ 4. $\frac{1}{12}$

Sometimes addition and multiplication of fractions are both used to determine a desired outcome.

Example 9

What is the probability of spinning a blue t-shirt or a green sweatshirt using the spinners below?

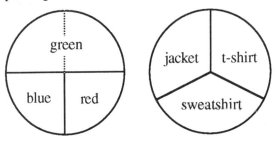

Step 1:   Determine the probabilities of each item:

P(blue t-shirt) = $\frac{1}{4} \cdot \frac{1}{3} = \frac{1}{12}$

P(green sweatshirt) = $\frac{1}{2} \cdot \frac{1}{3} = \frac{1}{6}$

Step 2:   Add the probabilities since either outcome is desired: $\frac{1}{12} + \frac{1}{6} = \frac{3}{12}$

The probability of getting either the blue t-shirt or the green sweatshirt is $\frac{3}{12}$ or $\frac{1}{4}$.

Problems

1.   Martha is spinning for her new prize. She wants either the blue stuffed bear or the stuffed pink flamingo. What is the probability she will get the prize she wants?

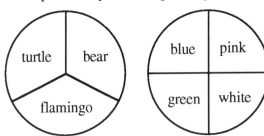

2. Carlos is playing a game with his friends. He can win either by rolling a 4 on a die and spinning blue or by rolling a 3 and spinning green. What is the chance that he will win the game?

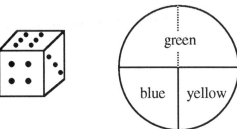

Answers

1. $\dfrac{2}{12} = \dfrac{1}{6}$

2. $P(4, \text{blue}) = \dfrac{1}{6} \cdot \dfrac{1}{4} = \dfrac{1}{24}$

$P(3, \text{green}) = \dfrac{1}{6} \cdot \dfrac{1}{2} = \dfrac{1}{12}$

$P(4, \text{blue or } 3, \text{green}) = \dfrac{1}{12} + \dfrac{1}{24} = \dfrac{3}{24} = \dfrac{1}{8}$

TREE DIAGRAMS

A tree diagram is a visual method to find possible outcomes or combinations (called "permutations" in mathematics). It is useful when other methods, such as an organized list, are cumbersome or there are more than two outcomes and a two-dimensional probability rectangle is not feasible. Students are introduced to tree diagrams in Year 1, Chapter 10, problem JM-62 on page 379 in a problem involving tossing three coins.

Tree diagrams are usually drawn horizontally. That is, each stage of the experiment is shown from left to right. Tree diagrams can be done with 2, 3, 4, or more possibilities.

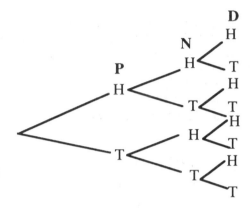

For example, suppose you are flipping a penny, a nickel, and a dime. Start with the penny and show the two possible outcomes (shown under the "P" at right). Since there were two possible outcomes for the penny, there are two possible outcomes for the nickel for each of the penny's outcomes. At this point, the tree diagram shows four possible outcomes for the flip of a penny followed by the flip of a nickel (under the "N").

Next flip the dime. Each flip is independent of the other coin flips. If you flip heads on the penny and heads on the nickel, you could get heads or tails on the dime. By following each branch in the diagram starting from the penny, you can see that there are eight possible outcomes when you flip three coins (shown under the "D"). A way to calculate that there are eight possible outcomes is to note that there are two outcomes for the penny, two for the nickel, and two for the dime, so 2(2)(2) = 8.

By following the branches from the start, far left, to the end of the branch, far right, you see all the possible outcomes: HHH, HHT, HTH, HTT, THH, THT, TTH, and TTT.

For additional information, see Year 1, Chapter 10, problem JM-62 on page 379.

Example 1

At a class picnic Will and Jeff were playing a game where they would shoot a free throw and then flip a coin. Each boy only makes one free throw out of three attempts. Use a tree diagram and a rectangular grid to complete parts (a) through (d) below.

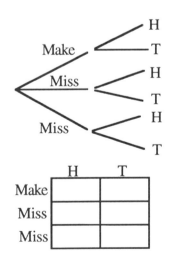

a)   P(miss, head) = ?

b)   P(miss, tail) = ?

c)   P(make, H) = ?

d)   P(make, T) = ?

By tracing the branches or counting the small rectangles, the probabilities are:

P(make, heads) = $\frac{1}{6}$, P(make, tails) = $\frac{1}{6}$, P(miss, head) $\frac{2}{6}$, and P(miss, tails) = $\frac{2}{6}$

Note that the sum of the probabilities in parts (a) through (d) is one.

Example 2

The local ice cream store has choices of plain, sugar, or waffle cones. Their ice cream choices are vanilla, chocolate, bubble gum, or frozen strawberry yogurt. Find all possible ice cream cone combinations.

There are 12 possible combinations.

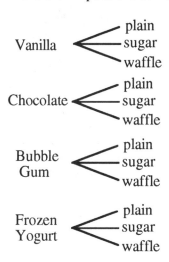

Example 3

Suppose the following toppings are available for the ice cream cones in example 2: sprinkles, chocolate pieces, and chopped nuts. There are now three more possible outcomes for <u>each</u> of the 12 outcomes in example 2, so 3(12) = 36 possible combinations.

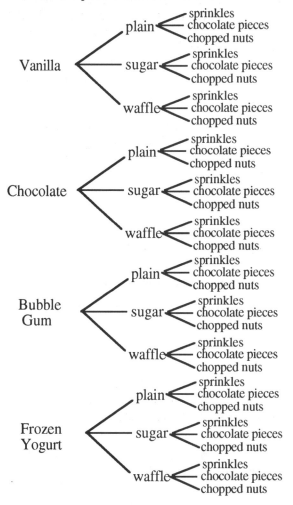

Problems

Draw tree diagrams to solve these problems.

1. How many different combinations are possible when buying a new bike if the following options are available:

    - mountain bike or racing bike
    - black, red, yellow, or blue paint
    - 3-speed, 5-speed, or 10-speed

ANSWER

There are 24 possible combinations as shown at right.

2. A new truck is available with:
    - standard or automatic transmission
    - 2-wheel or 4-wheel drive
    - regular or king cab
    - long or short bed

    How many combinations are possible?

ANSWER

There are 16 possible combinations as shown at right.

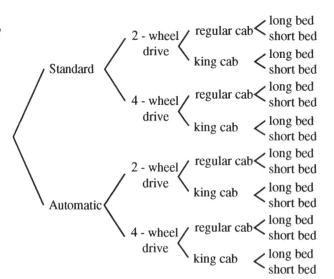

Statistics, Data Analysis, and Probability